Marine Biology: Biodiversity and Ecology

Marine Biology: Biodiversity and Ecology

Stefan Jenkins

Syrawood
PUBLISHING HOUSE

New York

Published by Syrawood Publishing House,
750 Third Avenue, 9th Floor,
New York, NY 10017, USA
www.syrawoodpublishinghouse.com

Marine Biology: Biodiversity and Ecology
Stefan Jenkins

© 2019 Syrawood Publishing House

International Standard Book Number: 978-1-68286-818-8 (Hardback)

Cataloging-in-Publication Data

Marine biology : biodiversity and ecology / Stefan Jenkins.
     p. cm.
Includes bibliographical references and index.
ISBN 978-1-68286-818-8
1. Marine biology. 2. Marine biodiversity. 3. Marine ecology. I. Jenkins, Stefan.
QH91 .M37 2019
578.77--dc23

TABLE OF CONTENTS

Permissions

Index

PREFACE

The scientific study of marine organisms and marine life in marine habitats is approached from the domain of marine biology. Marine habitats include the sea, coral reefs, seagrass meadows, kelp forests, thermal vents, etc. Marine organisms range in size from microscopic zooplankton and phytoplankton to massive cetaceans. In diversity, these organisms occupy all niches of organization, such as plants and algae, fungi, vertebrates and invertebrates. Some of the sub-disciplines of marine biology are ichthyology, phycology and invertebrate zoology. This book is a compilation of chapters that discuss the most vital concepts in the field of marine biology. The topics included herein are of utmost significance and bound to provide incredible insights to readers. Coherent flow of topics, student-friendly language and extensive use of examples make this book an invaluable source of knowledge.

A detailed account of the significant topics covered in this book is provided below:

Chapter 1- The systematic study of marine life including the organisms inhabiting the sea is under the domain of marine biology. This is an introductory chapter, which will discuss in brief the principles of marine biology and marine science.

Chapter 2- Marine organisms are the organisms that live in the sea. These include marine microbes, plants and algae, fungi, vertebrates and invertebrates. This chapter has been carefully written to provide an extensive understanding of the diverse marine life on Earth, such as marine vertebrates and invertebrates, aquatic plants and marine microorganisms.

Chapter 3- Marine habitats are either coastal or open ocean habitats. The aim of this chapter is to explore the varied types of marine habitats on Earth. Some of the topics elaborated in this chapter cover the diverse aspects of coastal habitats and pelagic zone for a comprehensive understanding of marine habitats.

Chapter 4- Marine ecosystems are large aquatic ecosystems that include salt marshes, lagoons, estuaries, coral reefs, the deep sea, etc. The study of marine ecosystems is within the scope of marine ecology. All the diverse aspects of marine ecology including aquatic and fresh water ecology have been carefully analyzed in this chapter.

Chapter 5- Marine microbiology is the science concerned with the study of microbes that live in saltwater ecosystems, such as the open ocean, estuaries, coastal water and marine surfaces and sediments. The interaction within different communities of microorganisms and their interactions with the environment are also studied under this field. This chapter closely examines the key concepts of marine microbiology, the diverse marine microbes present in such ecosystems such as marine eukaryotes, prokaryotes and viruses, etc.

Chapter 6- The conservation of biological and physical marine resources and ecosystem functions falls under marine biodiversity conservation. An understanding of marine biodiversity conservation follows from a study of marine biodiversity, the factors influencing biodiversity and the strategies that can be implemented to conserve it. This chapter covers all such vital aspects for a detailed understanding of this subject.

It gives me an immense pleasure to thank our entire team for their efforts. Finally in the end, I would like to thank my family and colleagues who have been a great source of inspiration and support.

Stefan Jenkins

Chapter 1

Introduction to Marine Biology

The systematic study of marine life including the organisms inhabiting the sea is under the domain of marine biology. This is an introductory chapter, which will discuss in brief the principles of marine biology and marine science.

Marine Science

The oceans are part of the thin, outer shell of the Earth and marine science is the study of this envelope, from the deep sea to shallow coastal oceans: their biology, chemistry, geology and physics together make marine science a richly inter-disciplinary science. The oceans are dynamic and vast, they contain most of the Earth's water and carbon and surface heat, and much of its biomass, but they do not operate alone. In conjunction with the atmosphere, continents and ice (cryosphere), they form a working machine, driven mostly by energy derived from the sun and the Earth's interior.

Marine Scientists focus their work on both practical or applied problems and basic scientific questions. The oceans provide both bounty and peril; they provide a wealth of food, a vast water supply reservoir and are the source of most of the heat and carbon used in our climate system, they are the source of ½ the oxygen our biosphere needs; they also spawn large storms and hurricanes, and transmit energy over great distances as tsunamis, all of which endanger coastal populations, which are a significant fraction of the Earth's total population.

Today's oceans are under great stress. Pressures from a growing human population are increasing, with fishing, shipping and recreation crowding into regions of abundant shell-fish and other active fisheries. Global climate change is being felt at all scales, from the coastal oceans to major ocean basins. These changes are causing a variety of phenomena, including increases in the bleaching of corals, increases in the frequency and extents of poisonous algal blooms (red tides), increases in ocean water oxygen depletion (hypoxia or anoxia), and changes in oceanic acidity, among many others.

People interested in marine science and technology will study the following:

- Climate change
- Marine aquaculture
- Oceanography
- Environmental (marine and coastal) management

- Fisheries management

- Marine ecotourism and environmental education

- Marine pollution control

- Marine conservation.

Marine Biology

Marine biology is the study of life in the oceans and other saltwater environments such as estuaries and wetlands. All plant and animal life forms are included from the microscopic picoplankton all the way to the majestic blue whale, the largest creature in the sea—and for that matter in the world.

Importance of Studying Marine Biology

Marine Biology vs. Biological Oceanography

The study of marine biology includes a wide variety of disciplines such as astronomy, biological oceanography, cellular biology, chemistry, ecology, geology, meteorology, molecular biology, physical oceanography and zoology and the new science of marine conservation biology draws on many longstanding scientific disciplines such as marine ecology, biogeography, zoology, botany, genetics, fisheries biology, anthropology, economics and law.

Like all scientific disciplines, the study of marine biology also follows the scientific method. The overriding goal in all of science is to find the truth. Although following the scientific method is not by any means a rigid process, research is usually conducted systematically and logically to narrow the inevitable margin of error that exists in any scientific study, and to avoid as much bias on behalf of the researcher as possible. The primary component of scientific research is characterization by observations. Hypotheses are then formulated and then tested based on a number of observations in order to determine the degree to which the hypothesis is a true statement and whether or not it can be accepted or rejected. Testing is then often done by experiments if hypotheses can produce predictions based on the initial observations.

The essential elements of the scientific method are iterations and recursions of the following four steps:

- Characterization (observation)

- Hypothesis (a theoretical, hypothetical explanation)

- Prediction (logical deduction from the hypothesis)

- Experiment (test of all of the above)

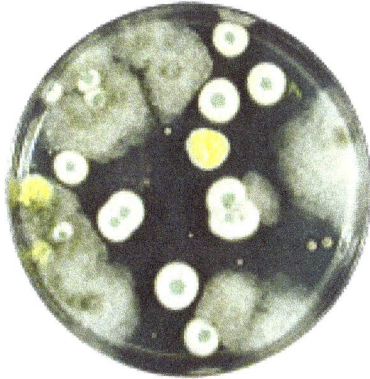

These steps are all used in the study of marine biology, which includes numerous sub fields including:

- Microbiology: The study of microorganisms, such as bacteria, viruses, protozoa and algae, is conducted for numerous reasons. One example is to understand what role microorganisms play in marine ecosystems. For example, bacteria are critical to the biological processes of the ocean, as they comprise 98% of the ocean's biomass, which is the total weight of all organisms in a given volume. Microbiology is also important to our understanding of the food chain that connects plants to herbivorous and carnivorous animals. The first level in the food chain is primary production, which occurs at the microbial level. This is an important biological activity to understand as primary production drives the entire food chain.

Scientists also study marine microbiology to find new organisms that may be used to help develop medicines and find cures for diseases and other health problems.

- Fisheries and Aquaculture: to protect biodiversity and to create sustainable seafood sources because of the world's dependence on fish for protein. There are many areas of study in this field.

  o The ecology of fisheries includes the study of their population dynamics, reproduction, behavior, food webs, and habitat.

  o Fisheries management includes studies on the impact of overfishing, habitat destruction, pollution and toxin levels, and ways to increase populations for sustainability as seafood.

  o Aquaculture includes research on the development of individual organisms and their environment. The objective is most often to develop the knowledge needed to cultivate certain species in a designated area in open water or in captivity in order to meet consumer demand. Technological advances have enabled seafood "farms" to produce high-demand products that traditional commercial fisheries cannot meet. This is a controversial area however, and an issue that will become of greater importance as our fish stocks continue to decline.

- Environmental marine biology: includes the study of ocean health. It is important for scientists to determine the quality of the marine environment to ensure water quality is sufficient to sustain a healthy environment. Coastal environmental health is an important area of environmental marine biology so that scientists can determine the impact of coastal development on water quality for the safety of people visiting the beaches and to maintain a healthy marine environment. Pollutants, sediment, and runoff are all potential threats to marine health in coastal areas. Offshore marine environmental health is also studied. For example, an environmental biologist might be required to study the impact of an oil spill or other chemical hazard in the ocean. Environmental biologists also study Benthic environments on the ocean bottom in order to understand such issues as the chemical makeup of sediment, impact of erosion, and the impact of dredging ocean bottoms on the marine environment.

- Deep-sea ecology: advances in technology of equipment needed to explore the deep sea have opened the door to the study of this largely unknown space in the sea. The biological characteristics and processes in the deep-sea environment are of great interest to scientists. Research includes the study of deep ocean gases as an alternate energy source, how animals of the deep live in the dark, cold, high pressure environment, deep sea hydrothermal vents and the lush biological communities they support.

- Ichthyology: is the study of fishes, both salt and freshwater species. There are some 25,000+ species of fishes including: bony fishes, cartilaginous fishes, sharks, skates, rays, and jawless fishes. Ichthyologists study all aspects of fish from their classification, to their morphology, evolution, behavior, diversity, and ecology. Many ichthyologists are also involved in the field of aquaculture and fisheries.

- Marine mammology: This is the field of interest to most aspiring marine biologists. It is the study of cetaceans—families of whales and dolphins, and pinnipeds (seals, sea lions, and the walrus). Their behaviors, habitats, health, reproduction, and populations are all studied. These are some of the most fascinating creatures in the sea; therefore, this is an extremely competitive field, and difficult to break into because the competition for research funding is also quite heavy.

One area of research currently being conducted on whales is the impact of military sonar on their health and well-being. The scientific community believes that high frequency sound waves cause internal damage and bleeding in the brains of whales, yet the military denies this claim. Military sonar can also interfere with the animal's own use of sonar for communication and echolocation. More research is needed; however, in recent years science has proven the claims to be valid and the military has begun limiting its use of sonar in specific areas.

- Marine ethology: The behavior of marine animals is studied so that we understand the animals that share the planet with us. This is also an important field for help in understanding how to protect endangered species, or how to help species whose habitats are threatened by man or natural phenomena. The study of marine animal behavior usually falls under the category of ethology because most often marine species must be observed in their natural environment, although there are many marine species observed in controlled environments as well. Sharks are most often studied in their natural habitat for obvious reasons.

Reasons for Studying Marine Biology

Life in the sea has been a subject of fascination for thousands of years. One of the most important reasons for the study of sea life is simply to understand the world in which we live. The oceans cover 71% (and rising) of this world, and yet we have only scratched the surface when it comes to understanding them. Scientists estimate that no more than 5% of the oceans have been explored. Yet, we need to understand the marine environment that helps support life on this planet, for example:

- Health of the Oceans/Planet
  - Climate change
  - Pollution (toxicology, dumping, runoff, impact of recreation, blooms)
  - Coral reefs

- ◦ Invasive species
- Human Health
  - ◦ Air quality
  - ◦ Dissolution of carbon dioxide
- Sustainability and Biodiversity
  - ◦ Overfishing
  - ◦ Endangered species
  - ◦ Impacts on the food chain
- Research and Product Development
  - ◦ Pharmaceuticals
  - ◦ Biomedical applications
  - ◦ Alternate energy sources

Tools for Studying Marine Biology

Advances in technology have opened up the ocean to exploration from the shallows to the deep sea. New tools for marine research are being added to the list of tools that have been used for decades such as:

- Trawling - has been used in the past to collect marine specimens for study, except that trawling can be very damaging to delicate marine environments and it is difficult to collect samples discriminately. However when used in the midwater environment, trawls can be every effective at collecting samples of elusive species with a wide migratory range.

- Plankton nets - plankton nets have a very fine weave to catch microscopic organisms in seawater for study.

- Remotely operated vehicles (ROVs) - have been used underwater since the 1950s. ROVs are basically unmanned submarine robots with umbilical cables used to transmit data between the vehicle and researcher for remote operation in areas where diving is constrained by health or other hazards. ROVs are often fitted with video and still cameras as well as with mechanical tools for specimen retrieval and measurements.

- Underwater habitats - the National Oceanic and Atmospheric Administration (NOAA) operates Aquarius, a habitat, 20m beneath the surface where researchers can live and work underwater for extended periods.

- Fiber optics - Fiber optic observational equipment uses LED light (red light illumination) and low light cameras that do not disturb deep-sea life to capture the behaviors and characteristics of these creatures in their natural habitat.

- Satellites - are used to measure vast geographic ocean data such as the temperature and color of the ocean. Temperature data can provide information on a variety of ocean characteristics such as currents, cold upwelling, climate, and warm water currents such as the Gulf Stream. Satellites are also used for mapping marine areas such as coral reefs and for tracking marine life tagged with sensors to determine migratory patterns.

- Sounding - hydrophones, the microphone's counterpart, detect and record acoustic signals in the ocean. Sound data can be used to monitor waves, marine mammals, ships, and other ocean activities.

- Sonar - similar to sounding, sonar is used to find large objects in the water and to measure the ocean's depth (bathymetry). Sound waves last longer in water than in air, and are therefore useful to detect underwater echoes.

- Computers - sophisticated computer technology is used to collect, process, analyze, and display data from sensors placed in the marine environment to measure temperature, depth, navigation, salinity, and meteorological data. NOAA implemented computer technology aboard its research vessels to standardize the way this data is managed.

Marine Biology versus Biological Oceanography

The difference between the terms "marine biology" and "biological oceanography" is subtle, and the two are often used interchangeably. As mentioned above, marine biology is the study of marine species that live in the ocean and other salt-water environments. Biological oceanography also studies marine species, but in the context of oceanography. So a biological oceanographer might study the impact of cold upwelling on anchovy populations off the coast of South America, where a marine biologist might study the reproductive behavior of anchovies.

Marine Biologist

Marine biologists study life in the oceans, and sometimes the oceans themselves. They may investigate the behavior and physiological processes of marine species, or the diseases and environmental conditions that affect them. They may also assess the impacts of human activities on marine life. Many marine biologists work under job titles such as wildlife biologist, zoologist, fish and wildlife biologist, fisheries biologist, aquatic biologist, conservation biologist, and biological technician.

Research Areas of a Marine Biologist

Marine biologists study marine organisms in their natural habitats. They may investigate a population's behaviors or physiology. Or, they may assess the condition of habitats, and the effects of human activity on those animals and habitats.

Their research typically involves conducting species inventories, testing and monitoring sea creatures exposed to pollutants, collecting and testing ocean samples, preserving specimens and samples of unknown species and diseases, and mapping the distribution, ranges, or movements of marine populations.

In some cases, they may recommend alternative industrial practices to minimize negative effects on marine species and habitats. They may also communicate their findings and recommendations by writing reports and scientific journal articles.

Some marine biologists specialize in marine biotechnology. In other words, they investigate the adaptations and advantages of marine species and how they might be applied to industrial processes. For instance, one biotech company has mimicked the structure of shark skin to create doorknobs that germs and viruses such as MRSA can't attach to. This is a promising and interesting area of the field.

Functional Areas of a Marine Biologist

The majority of marine biologists work for state and federal government agencies. These positions typically offer greater job security and more opportunity for advancement. Many work at private research laboratories or consulting firms. Others work for aquariums, zoos, and museums, or become faculty members in academia. Some become high school science teachers.

Marine biologists may spend a significant amount of time outdoors when conducting research. Fieldwork often involves working on ships to locate, tag, and monitor marine animals and their movements, and to collect seawater samples.

While marine biologist careers vary significantly, at its basic level, this type of scientist specializes in the living organisms in bodies of water. Most marine biologists have an area of specialty - some study mammals, or fish, single celled organisms like plankton, or plants and coral. While duties do vary from job to job, but the list below includes job duties that one typically encounters as a marine biologist:

- Review research and literature relating to current discoveries in the field
- Collect field and control samples of biological samples and non-living media in order to perform analyses
- Research the behavior and relationships among organisms in the marine environment
- Analyze the evolution and distribution of organisms and their environment in the ocean
- Use and maintain instrumentation used to track organisms and measure the properties of the environment
- Analyze the diversity and health of various components of the marine environment

- Consult and work to rebuild damaged marine ecosystems
- Use computer modeling to build predictive data for the marine ecosystem
- Consult with stakeholders regarding programs to monitor pollution
- Advocate for and monitor environmental compliance
- Provide assistance to fisheries management and coast guard units as required

Senior Marine Biologists often have a more broad set of job responsibilities that focus on management and leadership tasks. These responsibilities often include:

- Navigate various agencies and regulations in order to effectively monitor the ecosystem in question
- Construct grant proposals to fund research and fieldwork
- Draft scientific papers reporting research findings
- Facilitate a positive and challenging team environment with clear communication and mentoring opportunities
- Present research findings at conferences, and to policymakers and stakeholders
- Communicate with the public to help educate about issues affecting marine ecosystems like climate change and overfishing
- Determine jurisdictions for various laws and regulations
- Develop scope of work for projects as well as calculate project budgets and schedules Writes environmental assessments and impact statements
- Shares data with consideration to endangered species advocacy organizations

Jobs in this field are limited and competition is strong. Prominent research shows that job growth for zoologists and wildlife biologists as a whole will be 3-7% through 2022, slower than average for all occupations. Those with advanced math and computer skills will be at an advantage in the job market.

Chapter 2

Marine Life

Marine organisms are the organisms that live in the sea. These include marine microbes, plants and algae, fungi, vertebrates and invertebrates. This chapter has been carefully written to provide an extensive understanding of the diverse marine life on Earth, such as marine vertebrates and invertebrates, aquatic plants and marine microorganisms.

From the perspective of a land animal like us, the ocean can be a harsh environment. However, marine life is adapted to live in the ocean. Characteristics that help marine life thrive in a saltwater environment include the ability to regulate their salt intake or deal with large quantities of salt water, adaptations to obtain oxygen (e.g., a fish's gills), being able to withstand high water pressures, living in a place where they can get enough light, or being able to adjust to a lack of light. Animals and plants that live on the edge of the ocean, such as tide pool animals and plants, also need to deal with extremes in water temperature, sunlight, wind and waves.

Types of Marine Life

There is a huge diversity in marine species. Marine life can range from tiny, single-celled organisms to gigantic blue whales, which are the largest creatures on Earth. Below is a list of the major phyla, or taxonomic groups, of marine life.

Major Marine Phyla

The classification of marine organisms is always in flux. As scientists discover new species, learn more about the genetic makeup of organisms, and study museum specimens, they debate how organisms should be grouped. More information about the major groups of marine animals and plants is listed below.

Marine Animal Phyla

Some of the most well-known marine phyla are listed below. The marine phyla listed below are drawn from the list on the World Register of Marine Species.

- Annelida - this phylum contains segmented worms. An example of a segmented marine worm is the Christmas tree worm.

- Arthropoda - Arthropods have a segmented body, jointed legs and a hard exoskeleton for protection. This group includes lobsters and crabs.

- Chordata - Humans are in this phylum, which also includes marine mammals (cetaceans, pinnipeds, sirenians, sea otters, polar bears), fish, tunicates, seabirds and reptiles.

- Cnidaria - This is a diverse phylum of animals, many of whom have stinging structures called nematocysts. Animals in this phylum include corals, jellyfish, sea anemones, sea pens and hydras.

- Ctenophora - These are jelly-like animals, such as comb jellies, but they don't have stinging cells.

- Echinodermata - It includes such beautiful animals as sea stars, brittle stars, basket stars, sand dollars and sea urchins.

- Mollusca - This phylum includes snails, sea slugs, octopuses, squids, and bivalves such as clams, mussels and oysters.

- Porifera - This phylum includes sponges, which are living animals. They can be very colorful and come in a diverse array of shapes and sizes.

Marine Plant Phyla

There are also several phyla of marine plants. These include the Chlorophyta, or green algae, and the Rhodophyta, or red algae.

CENSUS OF MARINE LIFE

Abbreviation	CoML
Formation	2000
Purpose	Oceanography research
Website	coml.org

The Census of Marine Life was a 10-year, US $650 million scientific initiative, involving a global network of researchers in more than 80 nations, engaged to assess and explain the diversity, distribution, and abundance of life in the oceans. The world's first comprehensive Census of Marine Life — past, present, and future — was released in 2010 in London. Initially supported by funding from the Alfred P. Sloan Foundation, the project was successful in generating many times that initial investment in additional support and substantially increased the baselines of knowledge in often underexplored ocean realms, as well as engaging over 2,700 different researchers for the first time in a global collaborative community united in a common goal, and has been described as "one of the largest scientific collaborations ever conducted".

According to Jesse Ausubel, Senior Research Associate of the Program for the Human Environment of Rockefeller University and science advisor to the Alfred P. Sloan Foundation, the idea for a "Census of Marine Life" originated in conversations between himself and Dr. J. Frederick Grassle, an oceanographer and benthic ecology professor at Rutgers University. Grassle had been urged to talk with Ausubel by former colleagues at the Woods Hole Oceanographic Institution and was at that time unaware that Ausubel was also a program manager at the Alfred P. Sloan Foundation, funders of a number of other large scale "public good" science-based projects such as the Sloan Digital Sky Survey. Ausubel was instrumental in persuading the Foundation to fund a series of "feasibility workshops" over the period 1997-1998 into how the project might be conducted, one result of these workshops being the broadening of the initial concept from a "Census of the Fishes" into a comprehensive "Census of Marine Life". Results from these workshops, plus associated invited contributions, formed the basis of a special issue of *Oceanography* magazine; later that year, a workshop in Washington, D.C. addressed the formation of an Ocean Biogeographic Information System (OBIS) which would serve to collate existing knowledge about the distribution of organisms in the ocean and form the information management component of the Census.

The Census began in a formal sense with the announcement in May 2000 of eight grants totaling about 4 million US$ to create OBIS, as reported in *Science* magazine, 2 June. Meanwhile, an International Scientific Steering Committee was formed in 1999, which by 2001 envisaged "about half a dozen pilot [field] programs" for the period 2002-2004 which, along with OBIS and another project called "History of Marine Animal Populations" (HMAP), would provide the initial activities of the Census, to be followed by an additional series of field programs in 2005-2007, culminating in an analysis and integration phase in 2008-2010. During the operation of the Census, an additional non-field project was added, the Future of Marine Animal Populations (FMAP), which concentrated on forecasting the future of life in the oceans using modeling and simulation tools.

As a general method of working, project proposals would be debated within the Scientific Steering Committee and, if recommended for funding, a formal submission would be made to the Sloan Foundation for funding to support the Principal Investigators (PIs) and a Project Coordinator, meetings of project participants, and additional Synthesis and Education and Outreach activities. Since Sloan Foundation approval was dependent on promises of contributions from additional sources, and projects were encouraged to bring additional resources on board during their operation, the Foundation funds committed were effectively leveraged many times to provide a much more substantial program than would otherwise have been possible. As core infrastructure components, the Foundation also supported the Census' International Scientific Steering Committee and Secretariat, the U.S. National Committee, and an Education and Outreach Network to lift the project's visibility and engage other nations and organizations. The Census was ultimately estimated to have cost US $650 million, of which the Sloan Foundation contributed US $75 million with the remainder supplied by a large number of participating institutions, countries, and national and international organizations in the form of both direct and in-kind contributions.

The Census had its inception in a visionary leader (Grassle) who was able to convince a small group of colleagues of the need for such a project and find a like-minded individual (Ausubel) who saw the opportunity for the Sloan Foundation to take a key role in bring the Census to fruition. This was not leadership that sought out problems to solve – it identified an issue that could not be addressed through conventional national funding mechanisms and could only be approached through a large-scale collaborative endeavor. The Sloan Foundation saw the opportunity to facilitate new science that would also contribute knowledge for wide societal benefit.

Census Program

The Census consisted of three major component themes organized around the questions:

1. What has lived in the oceans?

2. What does live in the oceans?

3. What will live in the oceans?

The largest component of the Census involved investigating what currently lives in the world's oceans through 14 field projects. Each sampled the biota in one of six realms of the global oceans using a range of technologies. These projects were as follows:

- Arctic Ocean: ArcOD (Arctic Ocean Diversity)

- Antarctic Ocean: CAML (Census of Antarctic Marine Life)

- Mid-Ocean Ridges: MAR-ECO (Mid-Atlantic Ridge Ecosystem Project)

- Vents and Seeps: ChEss (Biogeography of Deep-Water Chemosynthetic Ecosystems)

- Abyssal Plains: CeDAMar (Census of Diversity of Abyssal Marine Life)

- Seamounts: CenSeam (Global Census of Marine Life on Seamounts)

- Continental Margins: COMARGE (Continental Margin Ecosystems)

- Continental Shelves: POST (Pacific Ocean Shelf Tracking Project)

- Near Shore: NaGISA (Natural Geography in Shore Areas)

- Coral Reefs: CReefs (Census of Coral Reefs)

- Regional Ecosystems: GoMA (Gulf of Maine Program)

- Microbes: ICoMM (International Census of Marine Microbes)

- Zooplankton: CMarZ (Census of Marine Zooplankton)

- Top Predators: TOPP (Tagging of Pacific Predators)

These field projects were complemented by the three non-field Census projects, namely HMAP, FMAP and OBIS. A series of National and Regional Implementation Committees (NRICs) was also established to progress the involvement of particular countries and regions in Census activities. Towards the end of the project, additional teams were created for education and outreach, and mapping and visualization products, while a "synthesis" group coordinated the final outcomes (publications, etc.).

Outcomes

During its lifespan, the Census involved some 2,700 scientists from more 80 countries who spent 9,000 days at sea participating in more than 540 census-badged expeditions, as well as uncounted near shore sampling events. In addition to many thousands of records of previously known species, Census scientists found more than 6,000 marine species potentially new to science and had completed formal descriptions of 1,200 of them up to 2010. Census scientists visited many parts of the global ocean to learn more about species ranging in size from the blue whale to minute zooplankton and microbes (bacteria and viruses); sampled from the world's coldest regions to the warm tropics, from deep-sea hydrothermal vents to coastal ecosystems; tracked the movements of fish and interrogated historical records to learn what the ocean used to be like before the influence of humans; and employed forecasting methods to predict what may happen to ocean life in the future. One of the largest scientific collaborations ever conducted; by 2011 the Census had produced over 3,100 scientific papers and many thousands of other information products, with over 30 million species distribution records freely available via OBIS.

As well as its tangible scientific legacy, the Census was instrumental in building a global community of researchers, many of whom had never collaborated before until they were brought together under the auspices of the Census, and a new approach to collaborative research. As Ian Poiner, outgoing chair of the Census has said, "The Census changed our views on how things could be done. We shared our problems and we shared our solutions. A fragmented research community: Marine biodiversity researchers had few active coordinated national and international research programs and taxonomic research in particular was underfunded and scattered in disparate organizations. No culture of collaboration and data sharing: Unlike the oceanographic community, marine biology was characterized by small research projects leading to publications but there was little experience or willingness to openly collaborate and share data. No recognized open-access data portal for marine biodiversity data: Unlike the "physical science" oceanographic community, there was no recognized data depository or common standards for sharing marine biodiversity data.

The Census, by any statistical measure, was a great success. The large number of scientific papers published and still to be published by Census participants, alone, would be sufficient. Instead the Census has achieved truly global science in biology. It did not profess to provide a complete Census of life in the ocean in 2010 but it did substantially increase the baselines of knowledge in often underexplored ocean realms. From this knowledge base future research and surveys will add more data that can be shared through web-based services such as OBIS. From this we may be able to derive estimates of population diversity, distribution and abundance for selected groups of organisms or regions. Undoubtedly much research on marine biodiversity would have been carried out over the last decade without the Census. But it would have lacked the global reach and the access to data and technologies that made the Census unique.

In 2011, the Census Steering Committee received the International Cosmos Prize in recognition of its decade of international ocean research spanning multiple scientific disciplines.

Partnerships

The Census partnered with the Encyclopedia of Life in creating pages for marine species, and supplied marine material for DNA barcoding in the Barcode of Life project. Google and Census of Marine Life partnered on Google Earth 5.0. Ocean in Google Earth contains a layer devoted to the Census of Marine Life that allows users to follow scientists from the Census on expeditions and see marine life and features found during the Census. A partnership with the French film company Galatée Films resulted in the production of the film *Oceans* which was released in 2009, featuring film of over 200 species at more than 50 global locations.

Marine Microorganism

The microbial world accounted for all known life forms for nearly 50 to 90% of Earth's history - Life itself likely began in the ocean. Marine microbes play many important roles in the Earth system: they influence our climate, are the major primary producers in the ocean, dictate much of the flow of marine energy and nutrients, and provide us with a source of medicines and natural products. Microbes (which include Bacteria, Archaea, microbial eukaryotes and their associated viruses) are as varied as the marine environments they come from. From the ice-covered polar regions of the Arctic and Antarctic to the boiling hydrothermal vents in the depths of the sea and the calcareous oozes made up of the skeletons of single-celled eukaryotes called foraminifera that once recorded the temperature of the sea surrounding them—microbes are everywhere and they help to shape the features of our planet, past and present. When it comes to exploring marine microbial life, we are still very much in the age of discovery. The Census of Marine Life is dedicated to enhancing our knowledge of this unknown microbial world through the development of an International Census of Marine Microbes.

Marine microbes are tiny, single-celled organisms that live in the ocean and account for more than 98 percent of ocean biomass.

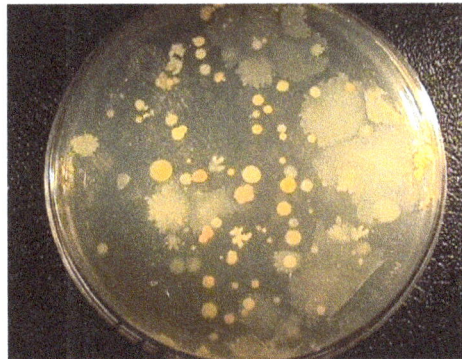

Marine bacteria growing on an agar plate.

The term "marine microbe" covers a diversity of microorganisms, including microalgae, bacteria and archaea, protozoa fungi, and viruses. These organisms are exceedingly small—only 1/8000th the volume of a human cell and spanning about 1/100th the diameter of a human hair. Up to a million of them live in just one milliliter of seawater.

Despite being found everywhere on Earth—in your belly, dirt, the oceans—the diversity and distribution of microorganisms remain under-sampled and uncharted, especially in the deep sea.

We know that microbes are the Earth's processing factories of biological, geological, and chemical (biogeochemical) interactions. Most marine microbes exist in highly organized and interactive communities that are versatile, complex, and difficult to analyze. They account for more than 98 percent of ocean biomass and possess as much variability as the environments they inhabit.

These organisms are capable of existing in practically any environment and garnering energy from a variety of sources, ranging from solar radiation to chemosynthesis. They play many different roles in the marine environment, from being the base of the food chain to controlling much of the flow of marine energy and nutrients and being essential to the ocean's health.

Marine microbe communities can evolve rapidly in response to environmental shifts and could be used as indicators of ocean change. In fact, marine microbes are "the canary in the coal mine" for the marine environment. In addition, they are also drivers of change in the ocean. Consequently, it is very important to acquire baseline information about these microorganisms against which future changes could be identified.

Marine Vertebrates

Marine vertebrates, classified under the Kingdom Animalia, Phylum Chordata and Subphylum Vertebrata, are among the most structurally complex organisms. The four main marine super classes and classes in Vertebrata are:

- Agnatha (jawless fishes)
    o Class Cephalaspidomorphi (lampreys)
    o Class Myxini (hagfishes)
- Gnathostomata
- Pisces (formerly Osteichthyes (now a synonym)) (bony fishes)
    o Class Actinopterygii (ray-finned fishes)
    o Class Elasmobranchii (formerly Chondrichthyes) (cartilaginous fishes: sharks, rays and skates)
    o Class Holocephali (cartilaginous fishes: chimaeras, ratfishes)
    o Class Sarcopterygii (lobe-finned fishes: lungfishes, coelacanths)
- Tetrapoda
    o Class Aves (birds)
    o Class Mammalia (mammals)
    o Class Reptilia (reptiles)

Super classes Agnatha and Pisces are all a form of fish. Superclass Agnatha contains the 105 species of jawless fish such as lampreys and hagfish.

Superclass Pisces contains at least 27,712 species of bony fishes including Class Actinopterygii which contains well over 20,000 species of ray-finned fishes, Class Elasmobranchii with at least 928 species of sharks, rays and skates, Class Holocephali with at least 928 species of chimaeras, and Class Sarcopterygii which contains the lobe-finned fishes such as lungfishes and coelacanths.

Under Superclass Tetrapoda, important marine classes include Class Aves (the birds) containing at least 9,842 species, Class Mammalia (all mammals) with at least 4,835 species, and Class Reptilia (the reptiles) which contains at least 3,082 species, which are all have species common in both terrestrial and marine environments.

Superclass Agnatha

This superclass includes the "primitive" jawless fish such as lampreys and hagfishes in the Classes Cephalaspidomorphi (lampreys), Myxini (hagfishes), and Pteraspidomorphi (fossil jawless vertebrates).

Class Actinopterygii

There are tens of thousands of species of bony fishes found in both marine and freshwater environments. Note that the plural form of fish ("fishes") is used when referring to more than one species of fish. Bony fishes come in all shapes and sizes and live in all marine zones. They range in size from the Bluefin tuna that measure up to 3 m to the stout infant fish, which measures about 7 mm. Most fish are slightly endothermic, meaning that they are able to regulate the temperature of their bodies, but not to the extent seen with mammals.

Endothermic bony fishes belong to the Suborder Scombroidei which includes about 122 species such as albacores, bonitos, cutlassfishes, frostfishes, hairtails, kingfishes, scabbardfishes, seerfishes, tuna, and wahoo. Endothermy uses a lot of energy but results in greater muscle control, better nerve signals, and improved digestion.

Class Elasmobranchii

This class consists of cartilaginous fish such as sharks, rays, and skates, whose skeletal structures are made up primarily of cartilage. This class contains some of the first marine species to develop paired fins. They all have 5-7 gill slits and lack swim bladders. Their large livers hold a large volume of oil that aids in buoyancy.

Because they lack a swim bladder, most sharks must swim constantly to avoid sinking. Constant swimming also aids in maintaining the flow of water and oxygen over the gills through their mouths. Their skin is covered in denticles rather than scales giving it a rough sandpaper-like feel. They reproduce through internal fertilization.

Class Holocephali

This class contains cartilaginous fishes such as chimaeras and ratfishes.

Class Sarcopterygii

This class contains the lobe-finned fishes such as the lungfishes and the recently discovered coelacanths.

Marine Birds (Class Aves)

Marine birds are characterized by a variety of adaptations to marine conditions. These adaptations include the ability of feathers to resist water, the presence of salt glands, curved bills, and webbed feet. Many marine birds are able to dive into the water to capture prey and some birds, like penguins, are able to swim into deeper water.

The preening gland releases waxes and water repellant fats that create a protective shield on the bird to prevent water from saturating the feathers. This protective shield also keeps the bird insulated and, when combined with feathers made out of keratin, the birds are essentially waterproof. Salt glands allow marine birds to drink salt water and expel the excess salt from their bodies. Salt glands work by condensing salt from the blood into the sinuses allowing the bird to sneeze out the excess. Some marine birds push out salt directly from salt glands.

Seabirds have a wide variety of eating mechanisms. Cormorants and Anhingas use their curved bills to puncture fish. Many marine birds dig into the sand for prey. Some have a distensible pouch such as those found in pelicans, frigatebirds, and cormorants. This pouch is located between the two parts of the lower mandible and is used to drain sea water from around the fish before it is eaten. Frigatebirds are known to snatch fish from other birds. Flamingos have a beak that can filter small invertebrates, algae, and other small organisms from water; their long bills and long legs enable them to stand in shallow water while they search for food.

Many birds will fly directly above the surface and look for fish swimming below. Black skimmers fly close to the surface and pluck fish near the surface of the water. Gulls and terns fly higher to find fish and plummet from the air to snatch them out of the water. Penguins are an unusual but very familiar marine bird that dives down into the ocean to look for fish.

As well as having a multitude of hunting strategies and bill adaptations, marine birds also have many different types of legs and feet. If a bird has very short legs with webbed feet, it is probably a very powerful swimmer. Webbed feet are referred to as totipalmate feet (having webbing connecting all four toes) and partially webbed feet are referred to as palmate feet. Birds that do a lot of swimming often rely upon what is called countercurrent exchange to keep the cold blood flowing up and away from their feet and from shocking their body. The arteries in these birds run very close to veins so that they are warm enough to allow the bird to swim in very cold water. Many marine birds also have very well developed eyesight and an amazing sense of smell. Birds in the Order Procellariiformes, called the "tubenoses" (albatrosses (13 species), diving-petrels (4 species), shearwaters and petrels (66 species), and storm-petrels (21 species)) can smell food up to 30 km away.

Marine birds are important to ecosystems for many reasons including their ability to move seeds throughout the environment and their predatory roles. They are also some of the most well-known and loved creatures of the sea. From the folkloric albatross to the march of the penguins, sea birds have fascinated humans for generations. However, they are threatened greatly by loss of habitat, oil spills, and entanglement in fishing nets or garbage, and the disruption in migration due to global warming.

Marine Mammals (Class Mammalia)

Marine mammals are vertebrates that have hair or fur, blubber, are warm blooded, use lungs to breathe air, bear live young, and produce milk through mammary glands. Marine mammals are very similar to land mammals with the exception of a thick layer of blubber instead of thick fur for insulation. They also typically have long bodies, which allow them to move swiftly through the water. Although they breathe air for oxygen, they are able to stay underwater for long periods of time because of their ability to hold extra oxygen in their muscles and blood. Many marine mammals have an excess amount of blood and can direct it to the most important organs when necessary during deep dives. They can also slow their heartbeat for more efficient oxygen use when diving.

The four most common groups of marine mammals include:

- Pinnipeds (Family Otariidae (sea lions and fur seals), Family Phocidae (seals), Family Odobenidae (walruses));

- Carnivores (Pinnipeds and Family Ursidae (polar bears) and Family Mustelidae (sea otters));

- Cetaceans (Suborder Mysticeti (baleen whales) and Suborder Odontoceti (toothed whales such as sperm whales, orca, dolphins, and porpoises)); and

- Sirenians (Family Dugongidae (2 species of dugongs) and Family Trichechidae (3 species of manatees)).

Whales are divided into the baleen whales (Suborder Mysticeti) and the toothed whales (Suborder Odontoceti - sperm whales, dolphins, porpoises, beaked whales, belugas, narwhals, etc.). Both types of whales have highly developed senses, blubber to keep warm and long bodies enabling them to swim quickly. Baleen whales have baleen plates instead of teeth composed of rigid fibers that act like a filter to catch zooplankton and phytoplankton. The upper jaws of baleen whales are long and flat. Toothed whales such as sperm whales and dolphins use teeth to catch prey like fish, octopus, and squid. Dolphins have larger brains than porpoises and porpoises have more rounded rostrums, triangular-shaped dorsal fins and spade-like teeth.

There are three families of pinnipeds which include the Family Otariidae (sea lions and fur seals), Family Phocidae (true seals), and Family Odobenidae (walruses). Although pinnipeds have blubber, they are warm blooded. They can live on land or in water. Seals can live out at sea for months at a time only returning to land for the mating season and to molt. Seals are found all over the world in coastal areas. The fur seals and sea lions are part of the eared seal family Otariidae. These animals have large flippers in the front, tiny ear lobes and webbed back legs that can rotate around. For this reason, eared seals are capable of moving quite easily when in water or on land. With no ears, short flippers and no rotation of the back flippers, true seals of the Phocidae family are easily distinguishable from the eared seals mostly because they look very awkward when trying to move on land. Walruses have a little hair on their body, long tusks and short thick whiskers. They are also quite a bit larger than other pinnipeds.

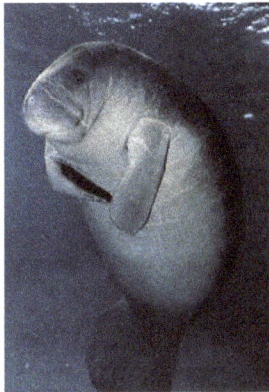

In the Order Sirenia, there are three species of manatees and two species of dugongs. Although dugongs have many of the characteristics of other marine mammals, they have a higher blubber ratio. Dugongs and manatees are gentle animals with large front flippers. Manatees are characterized by a round tail and can be found in Southeast U.S. coastal waters and in the coastal and inland waterways of Central America and along the northern coast of South America. Dugongs have a dolphin-like tail and are found discontinuously in coastal waters of east Africa from the Red Sea to northernmost South Africa, northeastern India, along the Malay Peninsula, around the northern coast of Australia to New Guinea and many of the island groups of the South Pacific.

Polar bears are the largest of all carnivores that live on land and are found throughout the arctic on sea ice, islands, and coastlines. These amazing predators often feed on ringed, bearded, harp, and hooded seals. The skin of polar bears is black with a layer of dense underfur and a layer of outer fur called guard hairs that are actually transparent. Polar bears are also protected from extreme temperatures with a thick (10 cm) layer of blubber.

Marine Reptiles (Class Reptilia)

Reptiles are vertebrates in the Class Reptilia, which includes four orders: Testudines (turtles, terrapins, and tortoises), Squamata (lizards, worm lizards, and snakes), Crocodilia (crocodiles, alligators, gavials, and caimans), and Rhynchocephalia (two species of lizard-like tuataras). The majority of marine reptiles are sea turtles and sea snakes, as well as the marine iguana, and the saltwater crocodile. In general a reptile is an animal that has very strong, dry skin sometimes covered with scales. Reptiles are cold-blooded or ectothermic, use lungs to breathe, and have tough skin without feathers or hair. Although reptiles are most commonly found in tropical and desert environment, they are also found in lakes, ponds, oceans, and even on top of mountains. Because they cannot regulate their internal body temperature, they are not found in extremely cold climates.

The more than 260 species of turtles and tortoises almost all have a protective shell surrounding their body. They range in size from the huge leatherback sea turtle (whose shell consists of bones beneath thick skin), which can reach up to 2.4 m in length and can weigh 907 kg to the smallest bog turtle that measures only 11.4 cm at most. Turtles that live in water have a lighter, flatter shell than the terrestrial species.

Lizards are the most common reptile with more than 2,700 species. Lizards usually have four legs with claws, however there are exceptions such as the worm lizard that usually has no appendages at all. Lizards range from a few inches to nearly 10 feet long, are typically insectivores, and inhabit trees, shrubs, or rocks. Some lizards will eat other vertebrates and others only eat plants. The Galapagos Islands are home to the marine iguana, the only iguana of the 416 known species that ventures into the ocean.

Snakes are basically reptiles with no appendages. There are about 2,000 species of snakes in the world measuring between 10 cm to 7.6 m in length. Snakes are thought to have evolved from lizards. Unlike other reptiles, snakes lack outside ears and have eyes covered with permanent transparent scales. Snakes have adapted to almost all habitats in the world and can also be found on lakeshores and even in ocean waters. Sea snakes have evolved to be extensively adapted to a fully aquatic life, except for the genus Laticauda, which retains ancestral characteristics that allows limited movement on land. Sea snakes are found in warm coastal waters from the Indian to the Pacific Oceans. All have paddle-like tails and most have laterally compressed bodies. However, unlike fish, they do not have gills and must come to the surface regularly to breathe. Nevertheless, they are among the most completely aquatic of all air-breathing vertebrates. Among this group are also species with some of the most potent venoms of all snakes. Most sea snakes will bite only when provoked, while others are much more aggressive at certain times of the year. Currently there are 62 species of sea snakes.

The Crocodilians are comprised of 23 species and include alligators, crocodiles, gavials, and caimans. Crocodilians can measure anywhere from 1.2 m to 6.2 m long in the saltwater crocodile. Crocodilians are usually found in tropical waters, although some alligators live in temperate climates especially in the US and China. Crocodilians have long flattened tails that enable them to swim efficiently through water. They breathe through nostrils located at the top of their head.

All reptiles have a sophisticated brain, central nervous system, and lungs. Some snakes only have one lung. Reptiles also have a three chambered heart with the exception of crocodiles, which have four chambers like mammals or birds. The digestive system of reptiles differs from other vertebrates in that waste, including that from the urinary system, the sexual organs, and the digestive system, empties into a holding tank called the cloaca. In the cloaca, water can be reabsorbed into the body to be used again.

The tough skin of reptiles helps protect the animal from desiccation or drying out. Many species use the tough skin as a form of protection from other animals and for protection during mating rituals. Some reptiles have the ability to change color for camouflage, communication, or sexual attraction. They reproduce through internal fertilization. Most reptiles lay eggs, although some lizards and snakes give birth to live young. Reptile eggs contain yolk and protein and are protected by a leathery or hard shell, which allows carbon dioxide and water to be exchanged while protecting the embryo from drying out or being consumed by bacteria. Sea turtles have been known to lay

150 eggs multiple times every season. Sea turtles lay their eggs in the sand and then leave them to hatch alone. The large number of eggs is necessary because many hatchlings do not survive. They are vulnerable to predation by birds, snakes, mammals, and sharks. Their nesting areas are also vulnerable to coastal development and eggs are vulnerable to human consumption. Sea turtles that do reach maturity can live up to 120 years in the wild. Alligators also have a long life span up to 70 years.

Fish

Fishes can have many forms. They can range in size from less than 1 cm (a goby) to over 12 m (a whale shark).

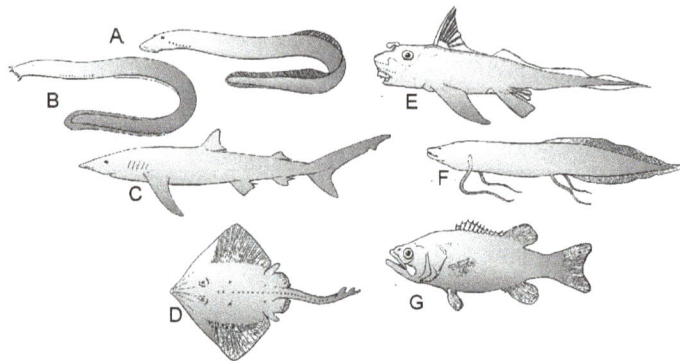

A. Lamprey   B. Hagfish   C. Shark   D. Ray   E. Chimaera   F. Lungfish   G. Teleost

They do not include seals, whales and dolphins (mammals), turtles and sea snakes (reptiles) or shrimps, lobsters, and crabs (crustaceans) or mussels (molluscs)

Some marine and freshwater invertebrates (animals with no backbone) and some animals with shells (such as molluscs) are called 'shellfish'. Therefore true fish may be called 'finfishes'.

Ichthyology is the study of the classification and the biology of fishes.

Fisheries science Is the study of the management and utilization of fish populations.

Main characteristics of fish:

- Animals that live in water.

- They have a skeleton inside the body – endoskeleton.

- They breathe through gills.

- They are cold-blooded which means that their bodies' temperature fluctuates with the temperature of the water around them.

- Most fishes lay eggs.

- They are usually covered with scales and swim using fins.

There are three main groups of fishes:

Jawless	Cartilagenous	Bony
Have no jaws or scales	Tiny "scales" are half buried in the skin and tooth like = denticles	Their bodies are usually covered with thin scales
Skeleton is made of cartilage	Skeleton of cartilage	Skeleton of bone
Several gill openings	Several gill openings	Have gill covers
e.g. hagfish and lampreys	e.g. sharks, rays and skates	e.g. snoek, yellowtail etc

Jawless Fishes

Although many ancient and now extinct fish were jawless, the only surviving ones today are the relatively insignificant hagfishes and lampreys.

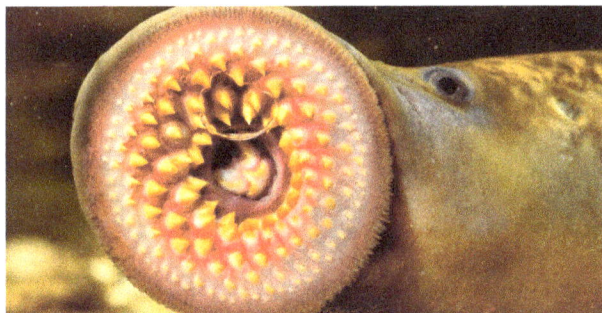

Hagfish

They have:

- Cartilaginous skeletons, eel-like bodies and no scales
- No eyes and no true teeth or vertebrae.
- The gills are carried on the one edge of the gill arches and are fused to the skull,
- They have fins and a multi-cuspid tongue-like structure (piston).
- They lack pectoral or pelvic fins.

Jawless fish are found in cool, temperate oceans in northern and southern hemispheres. There are about 83 species of jawless fish worldwide and 4 species in southern Africa.

Cartilaginous Fish

Cartilaginous fish evolved about 450 million years ago.

About 960 species of sharks are found in the oceans worldwide, about 300 species are found off southern African waters. Include sharks, rays, guitarfishes and sawfishes, known collectively as elasmobranchs and chimaeras.

The two main groups are:

- Elasmobranchii - Sharks, skates and rays have their upper jaw fused to their skull. Their teeth are usually separate and they have 5-7 pairs of gill openings/ gill slits. Rays have their gill openings located on the underside of the head and their pectoral fins are enlarged onto the typical flattened disc.

- Holocephali - Chimaeras have the upper jaw fused to the skull, teeth in the form of solid plates, and only one pair of external gill openings with a soft gill cover. The only living example in South African waters is the St Joseph shark also called the elephant fish.

Important features of cartilaginous fishes are:

- Skeleton is made of cartilage (found in our ears and nose). This is very soft and flexible material.

- Internal fertilization.

- 43% of cartilaginous fish (sharks and rays) lay eggs.

- All males have claspers for sperm transferral.

- Egg incubation periods vary from 1 to 15 months.

- Have true jaws and teeth and many species have rows of teeth, which move forward as front ones break off or wear out.

- Are primarily marine fish with some freshwater species.

- Sharks maintain a high concentration of urea in their blood and other tissues so as to maintain the ionic balance of their tissue relative to seawater. This prevents water loss.

- Sharks do not have scales, the skin is covered with tiny teeth like scales called denticles.

Most have a highly developed sense of smell and are acutely sensitive to vibration, which they sense through their lateral line allowing them to hunt their prey in very murky water where their sight is not effective.

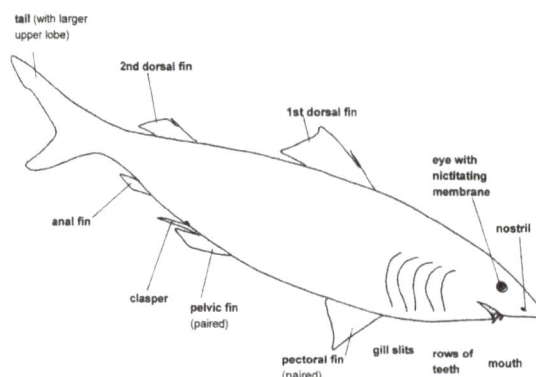

All cartilaginous fishes are carnivorous. Their prey can vary from large fish and marine mammals making-up the diet of the bigger pelagic sharks, to the mussels and crustaceans eaten by the more sluggish, bottom-dwelling rays and guitarfishes.

Feeding Methods

- The Great White sharks have triangular, heavily serrated teeth, suitable for cutting and tearing.

- Tiger shark teeth are also suitable for cutting and piercing.

- Ragged-tooth sharks' teeth are suitable for grasping and piercing.

- The rounded teeth of rays and guitarfish provide a mill-like grinding mechanism able to crush molluscs and crustaceans.

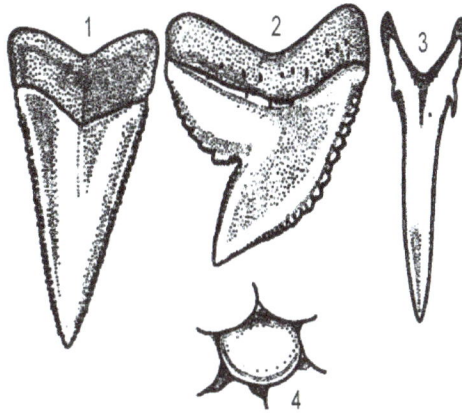

Different kinds of shark teeth

Working of the Jaws

The bottom jaw of many fast swimming sharks can hardly be seen as it fits so neatly under the top jaw, contributing to its streamlined profile. However when it is opened the upper jaw seems to dislocate, opening up the mouth to its fullest extent.

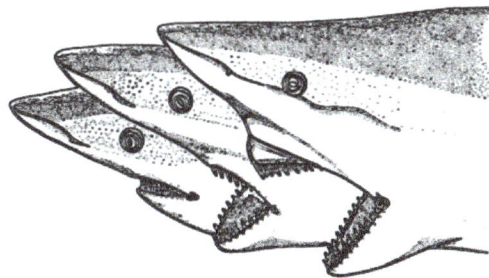

Jaw mechanism of a shark

Bottom dwelling sharks and rays have an opening behind their eye called a spiracle. The spiracle is used to allow flow of water over the gills, as the mouth of these animals is situated at the bottom in contact with sand and mud. If they were to take in water flow over their gills via their mouth the gills would be covered by sand or mud.

Stingrays	Skates
All stingrays have a flattened, disc-shaped body which is in most cases colorful and beautifully patterned.	Skates have the same expanded pectoral fins and disc shaped body. They are well camouflaged with the sandy or muddy habitat.
The enlarged pectoral fins are fused to the head.	
It has a long slender tail, dorsal and caudal fins are absent or reduced.. All stingrays have a pair of large serrated spines on top of the tail.	It has a long, thin tail with a row of enlarged denticles or thorns on the tail. It does not have a spine on the tail and will have two dorsal fins and a caudal fin at the tip of the tail.
They are bottom-living and range from estuarine shallows to depths of 1000 m.	They are also bottom dwelling but are not well represented in South African waters.
All are live bearing (Ovoviviparous) and in many species the embryos are nourished by unique uterine 'milk'.	Oviparous – eggs are laid in a tough flexible dark brown egg case called a mermaids purses.

Eagle Rays

The large body is equipped with enormous wing-like pectoral fins that effortlessly propel the ray through the water in a flying motion. Its diet consists of mussels and crustaceans, which are crushed by formidable rows of plate-like teeth. Up to four pups are born after a gestation period of one year.

Eagle ray

Reproduction in Sharks

Fish have three main methods of reproduction	
Oviparous	Producing eggs that develop and hatch outside the body of the female. E.g. pyjama catshark (mermaid's purse – egg case).
Ovoviviparous	Producing eggs that hatch within the body of the mother, but the embryos lack a placental attachment to the oviduct. e.g. Ragged tooth shark
Viviparous	Bringing forth living (active, free swimming) young. Gestation period can vary from 2 months to 2 years. Embryos are attached to mother in uterus with and umbilical cord. The mother feeds the babies through her own blood supply until they are ready to be born, much like a human baby. e.g. Great white shark

Summary of reproductive methods

Bony Fish

Bony spines or soft rays strengthen the fins of bony fishes. A single gill cover protects the gill chamber on either side of the body. The majority of bony fishes are completely covered with scales, with some exceptions like eels and barbels that completely lack scales.

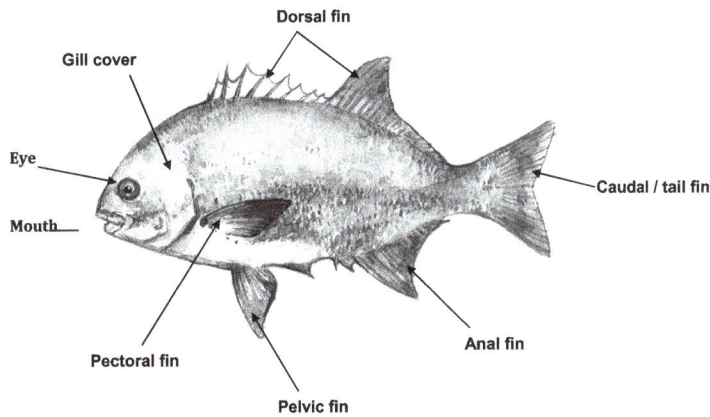

The external structure of a bony fish - Galjoen

The Body Shapes of Bony Fish are adapted to their environment and this illustration shows the terms used to describe their physical appearance.

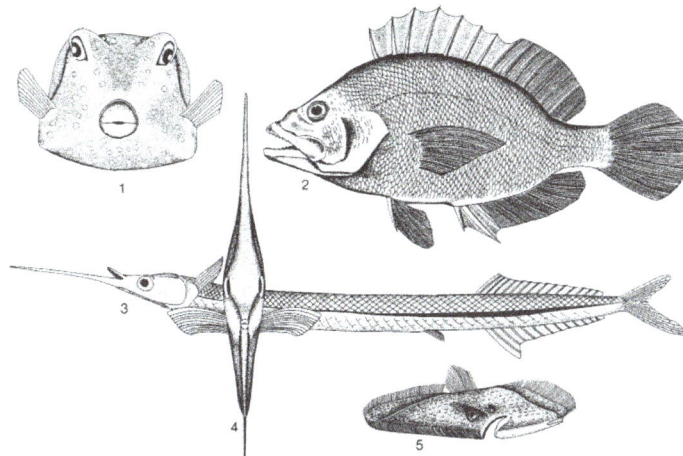

1. Square in cross-section e.g. boxfish and cowfishes.

2. Robust e.g. sea breams, rockcods.

3. Elongate, e.g. snoek and eels.

4. Compressed (that is flattened from side to side e.g. butterfly fish, moonfish, batfish.

5. Flat or depressed e.g. soles and flounder.

Swimming with Fins and Tails

Fins

Fishes can move in 3-D, forwards and backwards, left and right as well as up and down. They wiggle through the water with a series of S-shaped waves. Their dorsal and anal fins act like keels keeping them upright while the paired fins are used like oars to slow down or turn corners. A fin has a thin layer of skin supported by fin rays, which may have stiff and bony spines or soft and flexible (rays).

If the pelvic fins are set far back they are probably fast swimming species.

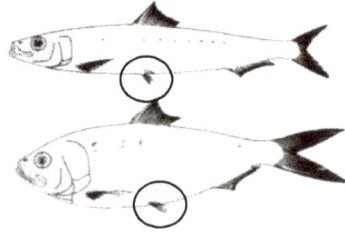

Slower moving fishes that live on reefs for example have large side fins, as they need to be able to manoeuvre around obstacles.

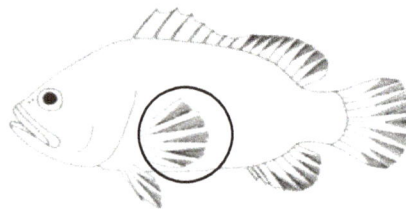

If the pelvic fins are placed well forward, sometimes even further forward than the pectorals, it is then typical of a bottom dwelling hovering species.

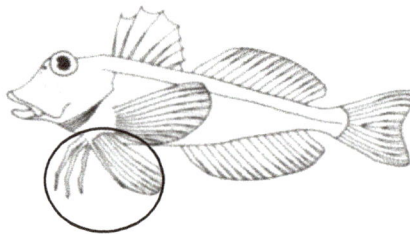

Tails

The tail or caudal fin is, in many fishes, the main means of forward propulsion and it uses the powerful muscles in the caudal peduncle with a sweeping side-to-side motion.

The shape of the tail and the general body shape is usually an indication of the fish's speed through the water.

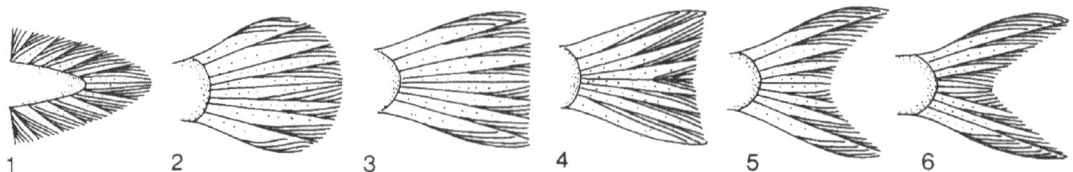

1.	Pointed
2.	Rounded
3.	Truncate
4.	Emarginate

| 5. | Lunate |
| 6. | Forked |

Faster, wide-ranging fish such as mackerel and tuna have forked tails, which are hydrodynamically more efficient than the rounded tails of the slow-moving fish such as kob and rockcods.

Scales and Slime

The majority of bony fishes are covered with thin transparent plates called scales.

Scales are in fact outgrowths of the skin and provide the animal with a flexible covering that, together with mucus (slime) secreted by the glands in the skin.

- Forms a barrier to bacteria and fungal infection
- Preventing the loss of body fluids (seals the body from osmoregulation).

Scales grow with the fish and the 'annual rings' are useful to determine a fish's age and history.

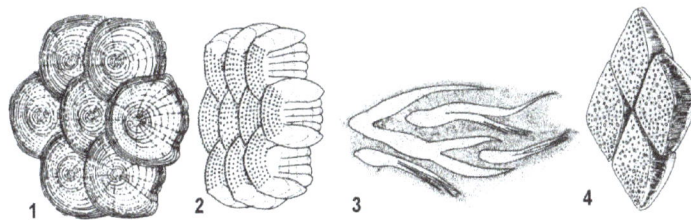

Cycloid (smooth edged) Ctenoid (rough edged) Lanceolate
(sharp and deeply embedded in the skin e.g. marlins), Rhomboid (not overlapping)

Fish Feeding

Fish like pilchards and sardines constantly sieve or strain plankton from the water passing through their gills. In contrast, certain predators may go without food for days or even weeks.

The shape of the mouth and the position of the mouth on the body give clues as to how it catches its food.

Head profiles showing typical mouth types

Elongate e.g. needlefishes - fast swimming surface carnivores;

1. Protractile (extended – 2a and retracted – 2b) e.g. purse mouths and John Dories – to suck up minute food items or to snatch relatively larger prey;

2. Superior or undershot – slow swimming predators, which ambush their prey e.g. rock cod and also fast swimming pelagic fish e.g. wolf herring;

3. Inferior or under slung e.g. Anchovies and bottom-dwelling barbels;

4. Terminal – characteristic of fish that nibbles or probes for food e.g. butterfly fishes, wrasses and grunters.

Snoek

Snoek have ferocious sets of teeth and are predators on fish and squid.

The feeding method of fish can be compared to certain "tools".

Mussel cracker

Vice

The mussel cracker has powerful jaws and large grinding teeth similar to a vice used for crushing shellfish and sea urchins.

Sea horse

The sea horse sucks in small food items like you would use a straw to drink your cooldrink.

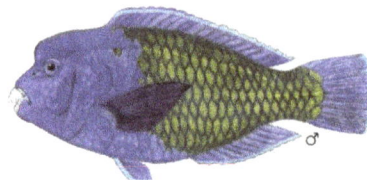
Parrotfish

Tweezers

The teeth of a parrotfish are fused into a sturdy beak for nibbling hard corals in the same way as you would use a tweezers.

The bones in a fish's head are not usually fused together so the fish can move and expand its head and jaws when feeding.

Some fish blow small prey out of the sand with their mouths similar to the way a bellow functions e.g. white steenbras.

White Steenbr

Fishes feed on a wide variety of foods, but the following major feeding groups are recognised.

Herbivores	Feeding mainly on plants - either phytoplankton, algae or higher plants.
Scavengers	Feeding on the remains of plants and animals, and associated organisms, usually on the bottom.
Detritivores	Feeding on detritus and associated diatoms and bacteria.
Predators	Feeding on other animals, including zooplankton, other fishes, and shellfishes. Worms etc.
Omnivores	Feeding on plant and animal food.

Growth Rates and Fish Survival

Fish Otoliths

Otoliths are the ear bones of fishes. These bones move in a pocket of sensory cells which record their position and give the fish information about balance, determine its orientation and speed of turning.

Information from otoliths has been used extensively to age fish and even those recovered from predators can provide an insight into the species and age of their prey. Obviously the big disadvantage is that only a dead fish can provide this information.

Finding the Age of Fish

Several methods have been explored using growth rings, much like those in a tree trunk. The principle is that certain substances are laid down at different rates under different conditions experienced seasonally, monthly or daily – which create alternating thick and thin layers seen as rings.

Otoliths are distinctive for different species of fish and have growth zones (e.g. summer and winter) which are laid down each year. These are best studied when a thin section (a slice) is taken from the otolith and polished.

Some fishes lay down two rings a year so the zones need to be verified against a fish with a known age from that species.

A chemical tetracycline is injected into a fish and will stain the ring being laid down at that time on the otolith. So, if the otolith is collected a year or two later, the number of growth rings laid down since the tetracycline staining can be counted.

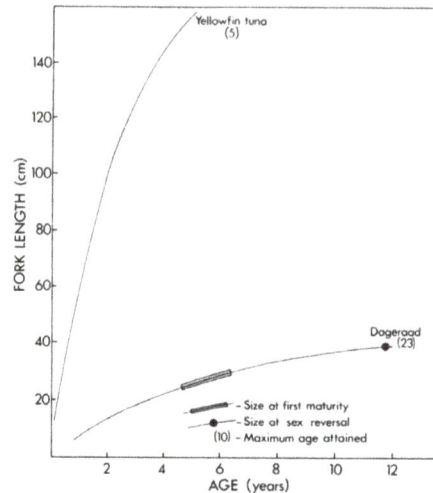

Graph showing the age of sexual maturity in
relation to the fork length of various species of fish

Growth Rates & Fish Survival

The rate at which fish grow is an important factor in determining how much fishing pressure their population can sustain. Fast-growing species (e.g. Yellowfin Tuna) tend to survive fishing pressure better than slow-growing species. Fast growing fish reproduce at a relatively young age (e. g Yellowfin Tuna is sexually mature at age 5), thus they produce young that will replace the older fish caught by the fishery.

Pelagic species, e.g. tuna and yellowtail, tend to be fast growing and can withstand relatively large fishing pressure.

In general, reef fish grow more slowly and reproduce later, so it is easier to over fish reef fish than pelagic species. Many reef fish (e.g. Dageraad and Red Roman) also undergo sex change during their lives. Fisheries tend to remove the big fish first, which results in the numbers of the physically larger sex being reduced. This reduces the reproductive potential of the population.

South African Sustainable Seafood Initiative (SASSI)

Marine resources must be wisely managed in order to ensure that we are able to use theme for as long as possible (sustainable use) and to conserve the coastal ecosystems and the diverse biological communities they support.

To accomplish this, we need both

- Effective fisheries regulations based on sound scientific information
- And the support of each individual.

Sustainable use of resources and conservation of the marine environment will only be achieved if consumers, seafood outlets and fishers support management programmes and comply with regulations.

To this end the South African Sustainable Seafood initiative (SASSI) has produced a Consumers Seafood pocket guide. Species or in some cases groups of similar species have been placed in different categories viz. Green, Orange and Red.

Each color has a different meaning viz.:-

Green: These are species that are from relatively healthy and well-managed populations that can sustain current fishing pressure. The fishing method used in catching these fish has little or no habitat damage and bycatch.

Orange: These species may be legally sold by registered commercial fishers and retailers.

However consumption of these species should be done with caution as: -

- The species may be rare due to overfishing

- The fishery may be damaging the environment through the method used

- The fishery has a high bycatch rate – accidental catch

Red: These species may be from unsustainable populations or are specially protected species and are by law illegal to buy or sell in South Africa.

Amphibians

Amphibians are animals that live in both water and on land. The word 'Amphibia' means dual lives. Amphibians are cold-blooded vertebrates that include the well-known frogs and toads. Being cold-blooded means that they depend on environmental sources of heat to regulate their body heat and temperature. The class of Amphibia is made up of more than 3,500 species, which include the various order of amphibians. Most amphibians begin their lives in water and eventually adapt to life on land by developing lungs and limbs that allow them to move on land. The larvae mature while in the water. At this young stage, the offspring breathe through the gills and after some time they develop lungs through a process known as metamorphosis. The class of Amphibia is made of three orders namely; Anura (toads and frogs), Urodela or Caudata (newts and salamanders), and Apoda or Gymnophiona (caecilians).

About 400 million years ago in the Devonian era, amphibians evolved from fish. The main reason for this evolutionary process was because of the rapid increase in the spread of dry land on earth. As a result, certain fish adapted to the changing conditions by developing limbs to crawl on land and lungs to breathe while out of water. The evolutionary process due to environmental change led to the rise of amphibians, the 'double-life' organisms. Amphibians also developed a backbone and became the first vertebrates to live on land. They returned to water for breeding purposes while feeding mostly on land. Between 340-230 million years ago, the planet experienced alternating periods of wet and dry conditions which allowed for the occurrence of the largest variety of amphibians. However, only a few groups of amphibians survived until the present age, which can be traced back to no further than 200 million years ago.

Major Characteristics of Amphibians

Amphibians have characteristics that cross-over between fish and reptiles. At the youngster age, most of them function like fish while as adults, they have different characteristics that allow them to live on land. They are cold-blooded animals which regulate their body heat and temperature depending on the external environment.

Amphibians have scale-less skin that is very delicate and moist. They live close to water sources in order to dampen their skin. The skin greatly helps in regulating the body temperature but also makes them vulnerable to dehydration. In high temperatures, dehydration will lead to death. This is the reason why amphibians live close to marshes, swamps and ponds and other freshwater bodies.

They breathe oxygen through the skin. The skin plays an important role in gas exchange and in absorption of water. This is despite them having lungs which function rather poorly under certain conditions. The skin, therefore, plays a double role of protecting and absorbing the water and oxygen.

Certain frogs such as the brightly colored poison-dart frog, have skin that contains poisonous glands. The poison is used as a defense mechanism which can easily kill any predator or prey. The poison from frogs has been used by Native American Indians hunters to coat the tips of their spears and arrows.

Three Orders of Amphibians

All amphibians are classified according to bodily characteristics of their legs and tails.

Anura

Anura is the largest order of living amphibians with over 3,000 different varieties. Toads and frogs fall under the order of Anura. This group lacks a tail and are characterized by long hind limbs that are adapted for swimming and leaping. Anura amphibians live in freshwater regions although some may be found in drier habitats. Frogs and toads are different in their body characteristic. Toads usually have shorter hind limbs and drier skin that appears warty, while frogs have a thin smooth skin and long hinder limbs. Anura amphibians feed on a variety of invertebrates such as insects. They can also feed on small mammals, birds, and fish.

Urodeles

Newts and salamanders fall under this category. The largest amphibian, the Japanese salamander, measures up to 1.5 meters while the smallest member of this order measures 10 centimeters in length. In this order, the tail is more pronounced than the limbs which are usually underdeveloped. Their preferred habitat is near water bodies and under moist soil and rocks. They mostly feed on insects and worms. Some species live in water, such as the genus Siren, while others burrow in the mud. They have lungs and external gills to aid in breathing.

Apoda

Apoda consists of about 205 species. They are shaped like worms, legless, and blind. They can be found in mud where they live, especially in the tropical soils of Africa and South America. They measure between 10 centimeters and 1 meter in length.

Amphibian Life Cycle

The life of amphibians begins in water where the female lays eggs that are externally fertilized. After the eggs hatch into tadpoles, they breathe through external gills. Tadpoles have flat tails that are used for swimming, and feed on aquatic vegetation. Eventually, through metamorphosis, they experience physical changes that make them adults. This includes the developing of lungs and elaborate limbs that aid them in movement on land.

Important Roles of Amphibians

Amphibians such as frogs are vital to the balance of ecosystem in which they inhabit, both as predators or prey. They feed on pests and insects thereby reducing the spread of diseases to agricultural plants. This indirectly benefits agriculture. In certain cultures around the world, frogs are viewed as a source of luck and are cherished as important symbols in society. In medical research, the skin of amphibians is being studied due to their ability to resist virus infections. This could eventually provide an advance in the treatment of virus diseases such as AIDS.

Main Threats to the Existence of Amphibians

Today, the number of amphibian species has continued to decline due to a variety of reasons. This includes the pollution of freshwater ecosystems which provide habitat for most of the species. Ultraviolet radiation has also affected the thriving of amphibians due to their fragile skin. Additionally, diseases such as the Chytrid fungus have depopulated many amphibian habitats. Many have been wiped out at a rapid rate to the extent that they are not even noticeable. The loss of amphibians affects the balance of the ecosystem which in turn affects other animal and plant species on the planet.

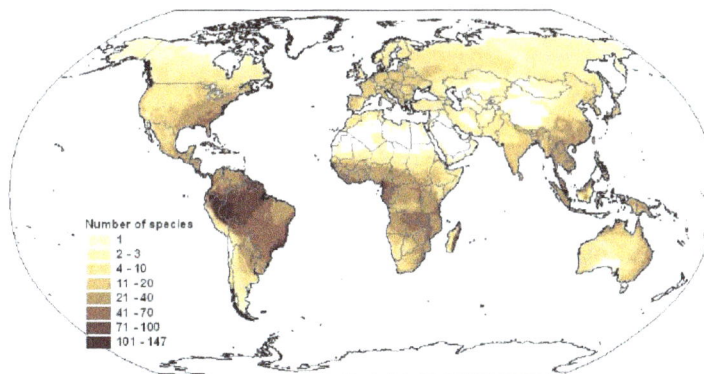

Global diversity of amphibians

Sea Bird

Seabirds are elusive and mysterious to most bird watchers because they are vastly different from more familiar songbirds and common backyard birds.

Types of Seabirds

Albatross

Albatrosses are tremendous pelagic birds. They are among the largest flying birds in the world, with mammoth wingspans and long, narrow wings that give them superb lift for easy flight. Unlike many marine birds, albatrosses also walk well on land. There are 22 albatross species in the world, all of which are part of the Diomedeidae bird family.

Auk

There are many types of auks, but with their compact bodies all are well adapted to northern seas and colder waters. Murres, puffins, and guillemots are all types of auks. These birds typically have upright posture on land but can be clumsy when walking. Most have black and white plumage, and many also have colorful bills or distinct markings. All auks are part of the Alcidae bird family.

Booby

Boobies are named for their somewhat silly or stupid expressions and history of gullible behavior where hungry sailors are concerned. These are beautiful tropical gannets with bright bills and feet, including the familiar and celebrated blue-footed booby. These are larger, heavy birds that can seem awkward in flight, but are often seen perched on buoys, cliffs, or rocks. All boobies are part of the Sulidae family.

Frigatebird

Also called pirate birds or man-of-war birds, frigatebirds are stately fliers with long, hooked bills, sharply pointed wings, and boldly forked tails. The red throat pouch of the males is distinctive, as is this bird's high, soaring flight. These are large birds that may circle slowly and elegantly before diving to steal fish from other birds. All five frigatebird species belong to the Fregatidae bird family.

Fulmar

Often confused with gulls because of their chunky build and short tails, fulmars are actually a type of petrel. They are opportunistic feeders and forage widely, including visiting land for trash or carrion. They are tubenoses with prominent bill structures that help filter salt out of the seawater they drink. There are only two fulmar species in the world, the northern fulmar and the southern fulmar. Both are included in the Procellariidae family.

Gannet

Gannets are large seabirds with chunky builds that appear somewhat ungainly on land, but they are strong fliers and powerful dive fishers. Their white plumage, buff-tinged head, and black wing-tips are distinctive for all three gannet species: northern gannet, cape gannet, and Australasian gannet. Because of those similarities, range is critical to tell the species apart. All three gannets are part of the Sulidae family along with their close relatives, the boobies.

Murre

A type of auk that resembles penguins, murres have dark and light counter shaded plumage and will swim underwater in pursuit of fish. When on land, they have an upright posture and waddling gait. There are only two murre species, the common murre and the thick-billed murre, both of which are found in northern oceans. These species are part of the Alcidae bird family with other types of auks, including puffins and guillemots.

Penguin

Flightless birds of the southern oceans, penguins are specialized seabirds well equipped for frigid waters with insulating plumage and fat. They are stunning swimmers and have specialized flippers rather than feathered wings. Despite their chilly reputation, however, several penguin species actually breed in tropical regions. The 18 penguin species vary greatly in size and range, though several types of penguins are physically similar. All penguins are in the Spheniscidae bird family.

Petrel

Petrels are small seabirds are distinguished by their elongated, tubular nostrils and their low, wave-skimming flight. They often stay so close to the water that they have the appearance of walking on its surface. Like many seabirds, petrels stay at sea nearly all their lives, returning to land only to breed. Birds called "petrel" are found in several scientific families. While some classifications are not universally accepted, the true petrels are generally considered part of the Procellariidae family.

Puffin

Puffins are large auks with broad, colorful bills and whimsical expressions. Because of this, they are often called sea clowns or sea parrots. They are powerful swimmers and nest in large breeding colonies, typically on offshore islands or isolated northern coasts. On land, they have an upright posture, and in flight their wing beats are rapid and their path direct. There are only three puffin species, horned, Atlantic, and tufted, and all are in the Alcidae bird family.

Shearwater

Shearwaters are small petrels with long wings that specialize in low, gliding flight. Their wingtips may brush the waves and "shear the water" as they fly, giving these birds their familiar name. There are more than 30 shearwater species and these birds can be found worldwide, but they often remain far out to sea. Where feeding conditions are ideal, they can gather in large groups, especially where upwellings bring prey such as fish, plankton, and squid close to the surface. Shearwaters are part of the Procellariidae family.

Tropicbird

Unlike many seabirds, tropicbirds have primarily white plumage and are easily identified by their very long streaming tail feathers. They have short legs and are shallow plunge divers when they hunt. There are only three species of tropicbird: the red-billed tropicbird, the white-tailed tropic-bird, and the red-tailed tropicbird. Because they are so distinct, they are the only members of the Phaethontidae family.

Evolution and Fossil Record

Seabirds, by virtue of living in a geologically depositional environment (that is, in the sea where sediments are readily laid down), are well represented in the fossil record. They are first known to occur in the Cretaceous period, the earliest being the Hesperornithiformes, like *Hesperornis regalis*, a flightless loon-like seabird that could dive in a fashion similar to grebes and loons (using its feet to move underwater) but had a beak filled with sharp teeth.

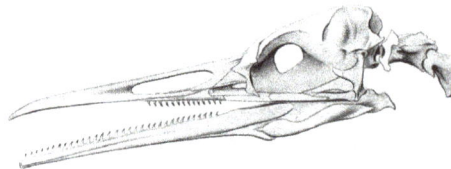

The Cretaceous seabird *Hesperornis*

While *Hesperornis* is not thought to have left descendants, the earliest modern seabirds also occurred in the Cretaceous, with a species called *Tytthostonyx glauconiticus*, which seems allied to the Procellariiformes and Pelecaniformes. In the Paleogene the seas were dominated by early Procellariidae, giant penguins and two extinct families, the Pelagornithidae and the Plotopteridae (a group of large seabirds that looked like the penguins). Modern genera began their wide radiation in the Miocene, although the genus *Puffinus* (which includes today's Manx shearwater and sooty shearwater) might date back to the Oligocene. The highest diversity of seabirds apparently existed during the Late Miocene and the Pliocene. At the end of the latter, the oceanic food web had undergone a period of upheaval due to extinction of considerable numbers of marine species; subsequently, the spread of marine mammals seems to have prevented seabirds from reaching their erstwhile diversity.

Characteristics

Adaptations to Life at Sea

Seabirds have made numerous adaptations to living on and feeding in the sea. Wing morphology has been shaped by the niche an individual species or family has evolved, so that looking at a wing's shape and loading can tell a scientist about its life feeding behavior. Longer wings and low wing loading are typical of more pelagic species, while diving species have shorter wings. Species such as the wandering albatross, which forage over huge areas of sea, have a reduced capacity for powered flight and are dependent on a type of gliding called dynamic soaring (where the wind deflected by waves provides lift) as well as slope soaring. Seabirds also almost always have webbed feet, to aid movement on the surface as well as assisting diving in some species. The Procellarii-formes are unusual among birds in having a strong sense of smell, which is used to find widely distributed food in a vast ocean, and possibly to locate their colonies.

Salt glands are used by seabirds to deal with the salt they ingest by drinking and feeding (particularly on crustaceans), and to help them osmoregulate. The excretions from these glands (which are positioned in the head of the birds, emerging from the nasal cavity) are almost pure sodium chloride.

Cormorants, like this double-crested cormorant, have plumage that is partly wettable. This functional adaptation balances the competing requirement for thermoregulation against that of the need to reduce buoyancy.

With the exception of the cormorants and some terns, and in common with most other birds, all seabirds have waterproof plumage. However, compared to land birds, they have far more feathers protecting their bodies. This dense plumage is better able to protect the bird from getting wet, and cold is kept out by a dense layer of down feathers. The cormorants possess a layer of unique feathers that retain a smaller layer of air (compared to other diving birds) but otherwise soak up water. This allows them to swim without fighting the buoyancy that retaining air in the feathers causes yet retain enough air to prevent the bird losing excessive heat through contact with water.

The plumage of most seabirds is less colorful than that of land birds, restricted in the main to variations of black, white or grey. A few species sport colorful plumes (such as the tropicbirds and some penguins), but most of the color in seabirds appears in the bills and legs. The plumage of seabirds is thought in many cases to be for camouflage, both defensive (the color of US Navy battleships is the same as that of Antarctic prions, and in both cases it reduces visibility at sea) and aggressive (the white underside possessed by many seabirds helps hide them from prey below). The usually black wing tips help prevent wear, as they contain melanins to make them black that helps the feathers resist abrasion.

Diet and Feeding

Seabirds evolved to exploit different food resources in the world's seas and oceans, and to a great extent, their physiology and behavior have been shaped by their diet. These evolutionary forces have often caused species in different families and even orders to evolve similar strategies and adaptations to the same problems, leading to remarkable convergent evolution, such as that between auks and penguins. There are four basic feeding strategies, or ecological guilds, for feeding at sea:

surface feeding, pursuit diving, plunge diving and predation of higher vertebrates; within these guilds there are multiple variations on the theme.

Surface Feeding

Many seabirds feed on the ocean's surface, as the action of marine currents often concentrates food such as krill, forage fish, squid or other prey items within reach of a dipped head.

Wilson's storm petrels pattering on the water's surface

Surface feeding itself can be broken up into two different approaches, surface feeding while flying (for example as practiced by gadfly petrels, frigatebirds and storm petrels), and surface feeding while swimming (examples of which are practiced by fulmars, gulls, many of the shearwaters and gadfly petrels). Surface feeders in flight include some of the most acrobatic of seabirds, which either snatch morsels from the water (as do frigate-birds and some terns), or "walk", pattering and hovering on the water's surface, as some of the storm petrels do. Many of these do not ever land in the water, and some, such as the frigatebirds, have difficulty getting airborne again should they do so. Another seabird family that does not land while feeding is the skimmer, which has a unique fishing method: flying along the surface with the lower mandible in the water—this shuts automatically when the bill touches something in the water. The skimmer's bill reflects its unusual lifestyle, with the lower mandible uniquely being longer than the upper one.

Surface feeders that swim often have unique bills as well, adapted for their specific prey. Prions have special bills with filters called lamellae to filter out plankton from mouthfuls of water, and many albatrosses and petrels have hooked bills to snatch fast-moving prey. Gulls have more generalized bills that reflect their more opportunistic lifestyle.

Pursuit Diving

The chinstrap penguin is a highly streamlined pursuit diver.

Pursuit diving exerts greater pressures (both evolutionary and physiological) on seabirds, but the reward is a greater area in which to feed than is available to surface feeders. Propulsion underwater can be provided by wings (as used by penguins, auks, diving petrels and some other species of petrel) or feet (as used by cormorants, grebes, loons and several types of fish-eating ducks). Wing-propelled divers are generally faster than foot-propelled divers. In both cases, the use of wings or feet for diving has limited their utility in other situations: loons and grebes walk with extreme difficulty (if at all), penguins cannot fly, and auks have sacrificed flight efficiency in favor of underwater diving. For example, the razorbill (an Atlantic auk) requires 64% more energy to fly than a petrel of equivalent size. Many shearwaters are intermediate between the two, having longer wings than typical wing-propelled divers but heavier wing loadings than the other surface-feeding procellariids, leaving them capable of diving to considerable depths while still being efficient long-distance travellers. The deepest diving exhibited by shearwaters is found in the short-tailed shearwater, which has been recorded diving below 70 m. Some albatross species are also capable of limited diving, with light-mantled sooty albatrosses holding the record at 12 m. Of all the wing-propelled pursuit divers, the most efficient in the air are the albatrosses, and they are also the poorest divers. This is the dominant guild in polar and subpolar environments, as it is energetically inefficient in warmer waters. With their poor flying ability, many wing-propelled pursuit divers are more limited in their foraging range than other guilds, especially during the breeding season when hungry chicks need regular feeding.

Plunge Diving

Gannets, boobies, tropicbirds, some terns and brown pelicans all engage in plunge diving, taking fast moving prey by diving into the water from flight. Plunge diving allows birds to use the energy from the momentum of the dive to combat natural buoyancy (caused by air trapped in plumage), and thus uses less energy than the dedicated pursuit divers, allowing them to utilize more widely distributed food resources, for example, in impoverished tropical seas. In general, this is the most specialized method of hunting employed by seabirds; other non-specialists (such as gulls and skuas) may employ it but do so with less skill and from lower heights. In brown pelicans the skills of plunge diving take several years to fully develop—once mature, they can dive from 20 m (70 ft) above the water's surface, shifting the body before impact to avoid injury. It has been suggested that plunge divers are restricted in their hunting grounds to clear waters that afford a view of their prey from the air, and while they are the dominant guild in the tropics, the link between plunge diving and water clarity is inconclusive. Some plunge divers (as well as some surface feeders) are dependent on dolphins and tuna to push shoaling fish up towards the surface.

Kleptoparasitism, Scavenging and Predation

This catchall category refers to other seabird strategies that involve the next trophic level up. Kleptoparasites are seabirds that make a part of their living stealing food of other seabirds. Most famously, frigatebirds and skuas engage in this behavior, although gulls, terns and other species will steal food opportunistically. The nocturnal nesting behaviour of some seabirds has been interpreted as arising due to pressure from this aerial piracy. Kleptoparasitism is not thought to play a significant part of the diet of any species, and is instead a supplement to food obtained by hunting. A study of great frigatebirds stealing from masked boobies estimated that the frigatebirds could at most obtain 40% of the food they needed, and on average obtained only 5%. Many species of gull will feed on seabird and sea mammal carrion when the opportunity arises, as will giant petrels. Some species of albatross also engage in scavenging: an analysis of regurgitated squid beaks has shown that many of the squid eaten are too large to have been caught alive, and include mid-water species likely to be beyond the reach of albatrosses. Some species will also feed on other seabirds; for example, gulls, skuas and pelicans will often take eggs, chicks and even small adult seabirds from nesting colonies, while the giant petrels can kill prey up to the size of small penguins and seal pups.

Life History

Seabirds' life histories are dramatically different from those of land birds. In general, they are K-selected, live much longer (anywhere between twenty and sixty years), delay breeding for longer (for up to ten years), and invest more effort into fewer young. Most species will only have one clutch a year, unless they lose the first (with a few exceptions, like the Cassin's auklet), and many species (like the tubenoses and sulids), only one egg a year.

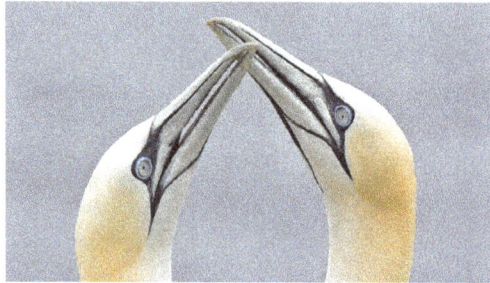

Northern gannet pair "billing" during courtship; like all seabirds except the phalaropes they maintain a pair bond throughout the breeding season.

Care of young is protracted, extending for as long as six months, among the longest for birds. For example, once common guillemot chicks fledge, they remain with the male parent for several months at sea. The frigatebirds have the longest period of parental care of any bird except a few raptors and the southern ground hornbill, with each chick fledging after four to six months and continued assistance after that for up to fourteen months. Due to the extended period of care, breeding occurs every two years rather than annually for some species. This life-history strategy has probably evolved both in response to the challenges of living at sea (collecting widely scattered prey items), the frequency of breeding failures due to unfavorable marine conditions, and the relative lack of predation compared to that of land-living birds.

Because of the greater investment in raising the young and because foraging for food may occur far from the nest site, in all seabird species except the phalaropes, both parents participate in caring for the young, and pairs are typically at least seasonally monogamous. Many species, such as gulls, auks and penguins, retain the same mate for several seasons, and many petrel species mate for life. Albatrosses and procellariids, which mate for life, take many years to form a pair bond before they breed, and the albatrosses have an elaborate breeding dance that is part of pair-bond formation.

Breeding and Colonies

Ninety-five percent of seabirds are colonial, and seabird colonies are among the largest bird colonies in the world, providing one of Earth's great wildlife spectacles. Colonies of over a million birds have been recorded, both in the tropics (such as Kiritimati in the Pacific) and in the polar latitudes (as in Antarctica). Seabird colonies occur exclusively for the purpose of breeding; non-breeding birds will only collect together outside the breeding season in areas where prey species are densely aggregated.

Common murres breed on densely packed
colonies on offshore rocks, islands and cliffs.

Seabird colonies are highly variable. Individual nesting sites can be widely spaced, as in an albatross colony, or densely packed as with a murre colony. In most seabird colonies, several different species will nest on the same colony, often exhibiting some niche separation. Seabirds can nest in trees (if any are available), on the ground (with or without nests), on cliffs, in burrows under the ground and in rocky crevices. Competition can be strong both within species and between species, with aggressive species such as sooty terns pushing less dominant species out of the most desirable nesting spaces. The tropical Bonin petrel nests during the winter to avoid competition with the more aggressive wedge-tailed shearwater. When the seasons overlap, the wedge-tailed shearwaters will kill young Bonin petrels in order to use their burrows.

Many seabirds show remarkable site fidelity, returning to the same burrow, nest or site for many years, and they will defend that site from rivals with great vigour. This increases breeding success, provides a place for returning mates to reunite, and reduces the costs of prospecting for a new site. Young adults breeding for the first time usually return to their natal colony, and often nest close to where they hatched. This tendency, known as philopatry, is so strong that a study of Laysan albatrosses found that the average distance between hatching site and the site where a bird established its own territory was 22 m; another study, this time on Cory's shearwaters nesting near Corsica, found that of nine out of 61 male chicks that returned to breed at their natal colony bred in the burrow they were raised in, and two actually bred with their own mother.

Colonies are usually situated on islands, cliffs or headlands, which land mammals have difficulty accessing. This is thought to provide protection to seabirds, which are often very clumsy on land. Coloniality often arises in types of bird that do not defend feeding territories (such as swifts, which have a very variable prey source); this may be a reason why it arises more frequently in seabirds. There are other possible advantages: colonies may act as information centres, where seabirds returning to the sea to forage can find out where prey is by studying returning individuals of the same species. There are disadvantages to colonial life, particularly the spread of disease. Colonies also attract the attention of predators, principally other birds, and many species attend their colonies nocturnally to avoid predation.

Migration

Like many birds, seabirds often migrate after the breeding season. Of these, the trip taken by the Arctic tern is the farthest of any bird, crossing the equator in order to spend the Austral summer in Antarctica. Other species also undertake trans-equatorial trips, both from the north to the south, and from south to north. The population of elegant terns, which nest off Baja California, splits after the breeding season with some birds travelling north to the Central Coast of California and some travelling as far south as Peru and Chile to feed in the Humboldt Current. The sooty shearwater undertakes an annual migration cycle that rivals that of the Arctic tern; birds that nest in New Zealand and Chile and spend the northern summer feeding in the North Pacific off Japan, Alaska and California, an annual round trip of 40,000 statute miles (64,000 km).

Pelican flock flying over HavanaBay area. These birds come to Cubaevery year from North America in the northern hemisphere winter season.

Arctic terns breed in the arctic and subarctic and winter in Antarctica.

Other species also migrate shorter distances away from the breeding sites, their distribution at sea determined by the availability of food. If oceanic conditions are unsuitable, seabirds will emigrate to more productive areas, sometimes permanently if the bird is young. After fledging, juvenile birds often disperse further than adults, and to different areas, so are commonly sighted far from a species' normal range. Some species, such as the auks, do not have a concerted migration effort, but drift southwards as the winter approaches. Other species, such as some of the storm petrels, diving petrels and cormorants, never disperse at all, staying near their breeding colonies year round.

Away From the Sea

While seabirds spend their lives on the ocean, many seabird families have many species that spend some or even most of their lives inland away from the sea. Most strikingly, many species breed tens, hundreds or even thousands of miles inland. Some of these species still return to the ocean to feed; for example, the snow petrel, the nests of which have been found 480 kilometres (300 mi) inland on the Antarctic mainland, are unlikely to find anything to eat around their breeding sites. The marbled murrelet nests inland in old growth forest, seeking huge conifers with large branches to nest on. Other species, such as the California gull, nest and feed inland on lakes, and then move to the coasts in the winter. Some cormorant, pelican, gull and tern species have individuals that never visit the sea at all, spending their lives on lakes, rivers, swamps and, in the case of some of the gulls, cities and agricultural land. In these cases it is thought that these terrestrial or freshwater birds evolved from marine ancestors. Some seabirds, principally those that nest in tundra, as skuas and phalaropes do, will migrate over land as well.

The more marine species, such as petrels, auks and gannets, are more restricted in their habits, but are occasionally seen inland as vagrants. This most commonly happens to young inexperienced birds, but can happen in great numbers to exhausted adults after large storms, an event known as a *wreck*, where they provide prized sightings for birders.

Relationship with Human

Seabirds and Fisheries

Seabirds have had a long association with both fisheries and sailors, and both have drawn benefits and disadvantages from the relationship.

Fishermen have traditionally used seabirds as indicators of both fish shoals, underwater banks that might indicate fish stocks, and of potential landfall. In fact, the known association of seabirds with land was instrumental in allowing the Polynesians to locate tiny landmasses in the Pacific. Seabirds have provided food for fishermen away from home, as well as bait. Famously, tethered cormorants have been used to catch fish directly. Indirectly, fisheries have also benefited from guano from colonies of seabirds acting as fertilizer for the surrounding seas.

Negative effects on fisheries are mostly restricted to raiding by birds on aquaculture, although long-lining fisheries also have to deal with bait stealing. There have been claims of prey depletion by seabirds of fishery stocks, and while there is some evidence of this, the effects of seabirds are considered smaller than that of marine mammals and predatory fish (like tuna).

Seabirds (mostly northern fulmars) flocking at a long-lining vessel

Some seabird species have benefited from fisheries, particularly from discarded fish and offal. These discards compose 30% of the food of seabirds in the North Sea, for example, and compose up to 70% of the total food of some seabird populations. This can have other impacts; for example, the spread of the northern fulmar through the United Kingdom is attributed in part to the availability of discards. Discards generally benefit surface feeders, such as gannets and petrels, to the detriment of pursuit divers like penguins.

Fisheries also have negative effects on seabirds, and these effects, particularly on the long-lived and slow-breeding albatrosses, are a source of increasing concern to conservationists. The bycatch of seabirds entangled in nets or hooked on fishing lines has had a big impact on seabird numbers; for example, an estimated 100,000 albatrosses are hooked and drown each year on tuna lines set out by long-line fisheries. Overall, many hundreds of thousands of birds are trapped and killed each year, a source of concern for some of the rarest species (for example, only about 2,000 short-tailed albatrosses are known to still exist). Seabirds are also thought to suffer when overfishing occurs.

Exploitation

The hunting of seabirds and the collecting of seabird eggs have contributed to the declines of many species, and the extinction of several, including the great auk and the spectacled cormorant. Seabirds have been hunted for food by coastal peoples throughout history—one of the earliest instances known is in southern Chile, where archaeological excavations in middens has shown hunting of albatrosses, cormorants and shearwaters from 5000 BP. This pressure has led to some species becoming extinct in many places; in particular, at least 20 species of an original 29 no longer breed on Easter Island. In the 19th century, the hunting of seabirds for fat deposits and feathers for the millinery trade reached industrial levels. Muttonbirding (harvesting shearwater chicks) developed as important industries in both New Zealand and Tasmania, and the name of one species, the providence petrel, is derived from its seemingly miraculous arrival on Norfolk Island where it provided a windfall for starving European settlers. In the Falkland Islands, hundreds of thousands of penguins were harvested for their oil each year. Seabird eggs have also long been an important source of food for sailors undertaking long sea voyages, as well as being taken when settlements grow in areas near a colony. Eggers from San Francisco took almost half a million eggs a year from the Farallon Islands in the mid-19th century, a period in the islands' history from which the seabird species are still recovering.

Both hunting and egging continue today, although not at the levels that occurred in the past, and generally in a more controlled manner. For example, the Māori of Stewart Island/Rakiura continue to harvest the chicks of the sooty shearwater as they have done for centuries, using traditional methods (called *kaitiakitanga*) to manage the harvest, but now work with the University of Otago in studying the populations. In Greenland, however, uncontrolled hunting is pushing many species into steep decline.

Other Threats

Other human factors have led to declines and even extinctions in seabird populations, colonies and species. Of these, perhaps the most serious are introduced species. Seabirds, breeding predominantly on small isolated islands, have lost many predator defence behaviors. Feral cats are capable of taking seabirds as large as albatrosses, and many introduced rodents, such as the Pacific rat, can take eggs hidden in burrows. Introduced goats, cattle, rabbits and other herbivores can lead to problems, particularly when species need vegetation to protect or shade their young. Disturbance of breeding colonies by humans is often a problem as well—visitors, even well meaning tourists, can flush brooding adults off a colony leaving chicks and eggs vulnerable to predators.

The build-up of toxins and pollutants in seabirds is also a concern. Seabirds, being apex predators, suffered from the ravages of DDT until it was banned; among other effects, DDT was implicated in embryo development problems and the skewed sex ratio of western gulls in southern California. Oil spills are also a threat to seabird species, as both a toxin and because the feathers of the birds become saturated by the oil, causing them to lose their waterproofing. Oil pollution threatens species with restricted ranges or already depressed populations.

This crested auklet was oiled in Alaska during the M/V Selendang Ayuspill

Conservation

The threats faced by seabirds have not gone unnoticed by scientists or the conservation movement. As early as 1903, U.S. President Theodore Roosevelt was convinced of the need to declare Pelican Island in Florida a National Wildlife Refuge to protect the bird colonies (including the nesting brown pelicans), and in 1909 he protected the Farallon Islands. Today many important seabird colonies are given some measure of protection, from Heron Island in Australia to Triangle Island in British Columbia.

Island restoration techniques, pioneered by New Zealand, enable the removal of exotic invaders from increasingly large islands. Feral cats have been removed from Ascension Island, Arctic foxes from many islands in the Aleutian Islands, and rats from Campbell Island. The removal of these introduced species has led to increases in numbers of species under pressure and even the return of extirpated ones. After the removal of cats from Ascension Island, seabirds began to nest there again for the first time in over a hundred years.

Seabird mortality caused by long-line fisheries can be greatly reduced by techniques such as setting long-line bait at night, dying the bait blue, setting the bait underwater, increasing the amount of weight on lines and by using bird scarers, and their deployment is increasingly required by many national fishing fleets. The international ban on the use of drift nets has also helped reduce the mortality of seabirds and other marine wildlife.

One of the Millennium Projects in the UK was the Scottish Seabird Centre, near the important bird sanctuaries on Bass Rock, Fidra and the surrounding islands. The area is home to huge colonies of gannets, puffins, skuas and other seabirds. The centre allows visitors to watch live video from the islands as well as learn about the threats the birds face and how we can protect them, and has helped to significantly raise the profile of seabird conservation in the UK. Seabird tourism can provide income for coastal communities as well as raise the profile of seabird conservation. For example, the northern royal albatross colony at Taiaroa Head in New Zealand attracts 40,000 visitors a year.

The plight of albatross and large seabirds, as well as other marine creatures, being taken as bycatch by long-line fisheries, has been addressed by a large number of non-governmental organizations (including BirdLife International, the American Bird Conservancy and the Royal Society for the Protection of Birds). This led to the Agreement on the Conservation of Albatrosses and Petrels, a legally binding treaty designed to protect these threatened species, which has been ratified by eleven countries as of 2008.

Depiction of a pelican with chicks on a stained glass window

Many seabirds are little studied and poorly known, due to living far out to sea and breeding in isolated colonies. However, some seabirds, particularly, the albatrosses and gulls, have broken into popular consciousness. The albatrosses have been described as "the most legendary of birds", and have a variety of myths and legends associated with them, and today it is widely considered unlucky to harm them, although the notion that sailors believed that is a myth that derives from Samuel Taylor Coleridge's famous poem, "The Rime of the Ancient Mariner", in which a sailor is punished for killing an albatross by having to wear its corpse around his neck.

Sailors did, however, consider it unlucky to touch a storm petrel, especially one that has landed on the ship.

Gulls are one of the most commonly seen seabirds, given their use of human-made habitats (such as cities and dumps) and their often fearless nature. They therefore also have made it into the popular consciousness – they have been used metaphorically, as in *Jonathan Livingston Seagull* by Richard Bach, or to denote a closeness to the sea, such as their use in *The Lord of the Rings* – both in the insignia of Gondor and therefore Númenor (used in the design of the films), and to call Legolas to (and across) the sea. Other species have also made an impact; pelicans have long been associated with mercy and altruism because of an early Western Christian myth that they split open their breast to feed their starving chicks.

Seabird Families

The following are the groups of birds normally classed as seabirds.

Sphenisciformes (Antarctic and southern waters; 16 species)

- Spheniscidae penguins

Procellariiformes (Tubenoses: pan-oceanic and pelagic; 93 species)

- Diomedeidae albatrosses

- Procellariidae fulmars, prions, shearwaters, gadfly and other petrels

- Pelacanoididae diving petrels

- Hydrobatidae storm petrels

Pelecaniformes (Worldwide; 8 species)

- Pelecanidae pelicans

Suliformes (Worldwide; about 56 species)

- Sulidae gannets and boobies

- Phalacrocoracidae cormorants

- Fregatidae frigatebirds

Phaethontiformes (Worldwide tropical seas; 3 species)

- Phaethontidae tropicbirds

Charadriiformes (Worldwide; 305 species, but only the families listed are classed as seabirds.)

- Stercorariidae skuas

- Laridae gulls

- Sternidae terns

- Rhynchopidae skimmers

- Alcidae auks

Marine Reptiles

Reptiles are the most diverse terrestrial vertebrates with about 12,000 described forms, including about 9,350 currently recognized species and about 3,000 subspecies.

About 260 million years ago reptiles evolved from aquatic amphibians and by the Jurassic (150–200 myr) modern reptiles had appeared. However, only a few reptile groups re-entered the oceans, primarily sea snakes (elapids related to cobras and kraits), and sea turtles. All major reptile groups, i.e. the snakes, lizards, turtles, and crocodiles, have at least a few members that enter marine habitats even though they may have never completely adapted to a life in the open sea. Here we give an overview of those reptiles that are found exclusively or at least occasionally in the oceans.

Sea Turtles

Sea turtles arose about 100 million years ago from terrestrial or fresh-water turtles. Currently only 7 species are extant although certain authors list Chelonia mydas agassizi as an eighth valid species. Sea turtles are found primarily along tropical coasts. However, some are also well known for their long journeys across the oceans. Most species nest along the coasts of Central and South America or in the Caribbean, although some species occasionally travel as far north as Scandinavia.

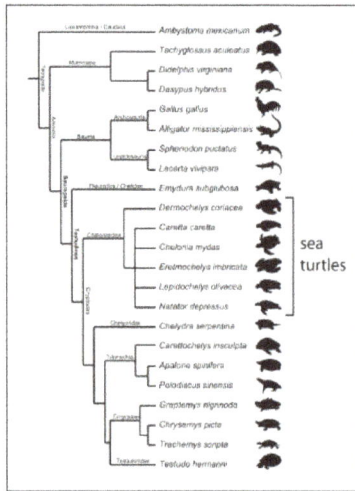

Phylogeny of sea turtles.
(A) Phylogenetic relationships of amniotes with the position of sea turtles relative to other vertebrates

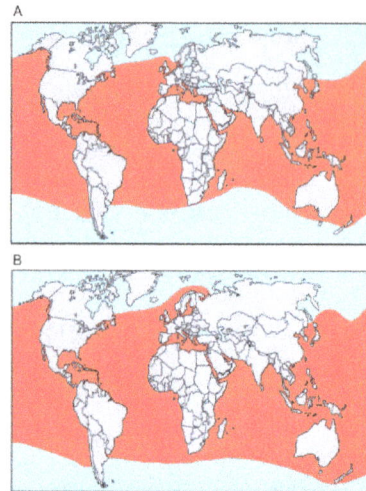

Distribution of sea turtles.
(A) Cheloniidae. (B) Dermochelyidae. Detailed maps for individual species.

Species	Common name	IUCN Red List status*
Caretta caretta	Loggerhead	Endangered
Chelonia mydas	Green Turtle	Endangered
Dermochelys coriacea	Leatherback	Critically Endangered
Eretmochelys imbricata	Hawksbill	Critically Endangered
Lepidochelys kempii	Atlantic Ridley	Critically Endangered
Lepidochelys olivacea	Pacific Ridley	Vulnerable
Natator depressus	Flatback Turtle	? (data deficient)

All species of sea turtles

Like many other reptiles, most if not all sea turtles seem to use temperature-dependent sex determination (TSD). TSD has been demonstrated for loggerhead (Caretta caretta), green (Chelonia mydas), leatherback (Dermochelys coriacea) and olive ridley (Lepidochelys olivacea) turtles. Since TSD is often sensitive to small changes in temperature, global warming may eventually affect sex ratios in these species and, together with their general vulnerabilities, have a dramatic effect on their reproductive rate and thus long-term survival.

Migration

Usually, sea turtles travel the seas on their own. However, during the nesting season they head towards their home beaches and may form large groups of turtles traveling together, even if they maintain distances of up to several hundred meters between individuals. The ability of sea turtles to find their original nesting sites has spurred considerable interest. Only recently it has been shown that sea turtles use geomagnetic sensing to orient themselves. While sea turtles appear to return to a certain geographic region, they not necessarily return to a specific. However, Kemp's Ridley does nest on only one location in Mexico, so it is especially endangered. It remains unclear though how precise other turtles are in returning.

Conservation

Most sea turtles are endangered or even critically endangered. The Olive Ridley Turtle (Lepido-chelys olivacea) is vulnerable and for the Flatback Turtle (Natator depressus) there is simply not enough data, although in previous years it had been classified as "vulnerable" too. The IUCN Red List says about the Green Turtle (Chelonia mydas):

"Analysis of historic and recent published accounts indicate extensive subpopulation declines in all major ocean basins over the last three generations as a result of overexploitation of eggs and adult females at nesting beaches, juveniles and adults in foraging areas, and, to a lesser extent, incidental mortality relating to marine fisheries and degradation of marine and nesting habitats. Analyses of subpopulation changes at 32 Index Sites distributed globally show a 48% to 67% decline in the number of mature females nesting annually over the last 3 generations."

Similar statements are true for all sea turtles: overexploitation and the destruction of nesting sites cause concern for these species, as tropical beaches are also in high demand for touristic reasons and easily exploited by locals.

Other Turtles Living in Brackish Environments

Of the more than 300 species of turtles only seven are truly marine while about 50 species are fully terrestrial, belonging to the family of tortoises, the Testudinidae. The majority of the remaining species, and most of the world's turtles, are aquatic or semi-aquatic freshwater species. However, a few are associated with estuarine and other brackish water habitats in an environment that is neither marine nor fresh water.

Genus	number of all species	brackish species
Malaclemys	1	1
Batagur	6	3
Carettochelys	1	1
Pelochelys	3	1

Turtles in brackish waters.

Included among the species that spend a portion or all of the year in estuarine habitats are the mangrove terrapins (Batagur affinis and B. baska) of south-east Asia and India, as well as the pig-nosed turtles (Carettochelys sp.) in southern New Guinea. Painted terrapins (Batagur borneoensis) of south-east Asia characteristically spend an even greater portion of their life cycle in estuarine and brackish waters, even laying their eggs on oceanfront beaches in the same areas as sea turtles. A few additional species of turtles, such as the giant softshell (Pelochelys cantori) of Asia, will enter estuarine brackish waters and even into full saltwater habitats temporarily, but the majority of their range includes freshwater habitats.

The only exclusively brackish water turtle in the world is the diamondback terrapin (Malaclemys terrapin), which is endemic to tidal creeks and salt marshes as well as the brackish portions of estuaries of the Atlantic and Gulf coasts in the United States, from Cape Cod to Texas. Diamondback terrapins in Florida are found in mangrove swamps. Two species of Batagur in Asia and the painted terrapin in Malaysia also occupy mangrove swamps. All of the species found in mangrove swamps also inhabit other brackish waters of coastal areas and most are associated to some degree with river estuaries, and the Asian species commonly enter freshwater systems as well. None of the turtle species in South America, Africa, or Europe has an affinity for brackish water conditions characteristic of the species noted above. However, many species of freshwater turtles that live in coastal areas around the world are known to enter brackish waters on occasion. The diamondback terrapin has a functional salt gland and can live indefinitely in fresh water or sea water. Other estuarine turtles have not been reported to have well developed salt glands comparable to those in the diamondback terrapin.

Malaclemys terrapin.
This species is the only terrestrial turtle with
significant adaptions to coastal habitats.

All turtle species that are brackish water inhabitants face severe conservation threats in all or part of their range because of the numerous environmental impacts on coastal systems throughout the world. Habitat degradation from pollution, sediment runoff, and other consequences of over-development as well as mortality as targeted commercial species or as bycatch from the fisheries industry take an increasingly greater toll on these species.

Marine Crocodiles

None of the currently 23 species of crocodile is truly marine (Crocodylus raninus has been re-validated as 24th species only recently but this has not been universally accepted). However, at least one species, the Saltwater or "Estuary" Crocodile (Crocodylus porosus) is regularly found in brackish waters of Southeast Asia and Australia. Other crocodiles have been found in tidal waters, such as Crocodylus johnstoni or C. acutus, but C. porosus is the only one that does show some adaptations to salt water.

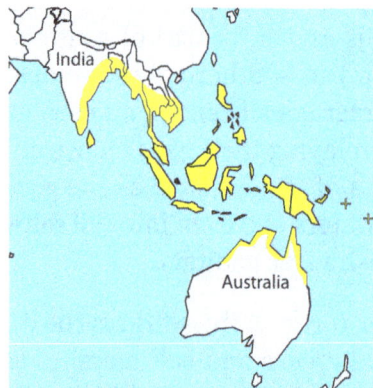

Distribution of the saltwater crocodile, Crocodylus porosus.
The range is shown in yellow. "+" symbols represent the Pacific islands that
are also inhabited by this species, including the Solomon islands and Vanuatu.

The saltwater crocodile is the largest living crocodile and thus the largest living reptile, reaching a total length of more than 6 meters. The species seems to be most closely related to C. siamensis, which is not particularly associated with brackish waters.

The saltwater crocodile has a high tolerance for salinity, being found in brackish water around coastal areas and in rivers. However, it is also present in freshwater rivers and swamps. Movement between different habitats occurs between the dry and wet season, and as a result of social status: juveniles are raised in freshwater areas, but eventually sub-adult crocodiles are usually forced out of these areas (used for breeding by dominant, territorial adults), into more marginal and saline areas. Subordinate animals unable to establish a territory in a tidal river system are either killed or forced out into the sea where they move around the coast in search of another river system. In recent years in northern Australia, saltwater crocodile populations in some areas have recovered to such an extent that increasing numbers are being forced further upstream into marginal habitat.

Crocodiles use their lingual salt glands to secrete excess salt ions. The morphology of these salt-secreting glands is highly conserved. These tissues are typified by their abundance of ion pumps, responsible for the maintenance of cellular electrochemical gradients through the movement of Na+ and K+ ions against their osmotic gradients. Crocodylus porosus possesses lingual salt glands, which function to remove excess Na+ and Cl− ions accumulated as a consequence of living in a marine environment. However, other crocodiles, such as the Nile crocodile, seem to have similar glands, even though they may be not as active or efficient as those in the saltwater crocodile. Such adaptations to seawater are evidence for a marine evolutionary origin of crocodiles.

Sea Snakes

Besides the sea turtles, the sea snakes are the reptiles that are best adapted to marine environments. The most typical feature of a sea snake is the vertically flattened paddle-like tail, which is not found in any other terrestrial or aquatic snakes. Sea snakes occur in the tropical and subtropical waters of the Indian and Pacific Oceans from the east coast of Africa to the Gulf of Panama. However, two specimens of Pelamis platurus reported from Namibia indicate that the species may be extending its range into the Atlantic Ocean. Most species are concentrated in the Indo-Malayan Archipelago, South China Sea, Indonesia and the Australian region. Sea snakes inhabit shallow waters along coasts and around islands, river mouths and up rivers for more than 150 km and they have also been found in lakes in Thailand, Cambodia, the Philippines and Rennell Island in the Solomon archipelago. However, information on precise geographical distribution and abundance for each species is still lacking.

A typical sea snake, Hydrophis belcheri.

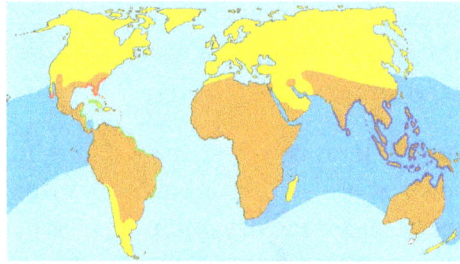

Distribution of marine snakes.

Terrestrial distribution represents terrestrial elapids (brown), marine distribution represents sea snakes, i.e. the subfamily Hydrophiinae of the Elapidae (blue). Dark blue: homalopsid snakes along the Asian and Australian coasts. Red: North-American Natricidae, green: neotropical Dipsadidae.

Notably, a few sea snakes are actually freshwater species, such as Hydrophis semperi, H. sibauensis, H. obscurus, and L. crockeri. However, all evidence indicates that these species have radiated into freshwaters independently from saltwater species.

Sea Snake Bite

Sea snake bite is the cause of human fatalities. The typical victim is a fisherman handling gape nets or down nets, sorting fish (and sea snakes) on board a trawling boat or dragging a net by wading in muddy coastal waters or in river-mouths. Some sea snakes are gentle, inoffensive creatures that bite only under provocation, but other species are much more aggressive (e.g. Aipysurus laevis, Astrotia stokesii, Enhydrina schistosa, Hydrophis elegans, H. macdowelli, H. major, H. ornatus and H. ocellatus). As a general rule all sea snakes must be handled with great caution: however, it is worth mentioning that when sea snakes do bite, they do not always inject much or any of their venom, so that only trivial severity of poisoning will be recognizable.

Taxonomy

The taxonomic status of the sea snakes is still under review and no general agreement exists at the moment. Traditionally sea snakes have been regarded as belonging to one family, Hydrophiidae, with Laticauda as the most primitive genus.

However, more recent results support the position that Laticaudinae and Hydrophiinae (true sea snakes) have evolved from different terrestrial elapids. The combined morphological and molecular results by Scanlon and Lee support that sea kraits (Laticauda) and Solomon Islands elapids are basal to the remaining Australian terrestrial elapids and true sea snakes (Hydrophiinae). The Australian Elapids and true sea snakes include three main lineages: a large-bodied oviparous lineage, a small-bodied oviparous lineage, and a viviparous lineage which also includes the true sea snakes. The results by Castoe et al. indicate that Laticauda is closest to some Asiatic elapids. However, these authors suggest to interpret these results cautiously because of long branch attraction. Sanders et al. indicate that Laticauda is the sister group to all other hydrophiines (oxyuranines), which is also consistent with a classic morphological feature: all oxyuranines have reduced the choanal process of the palatine and lost the lateral process, permitting novel jaw movements.

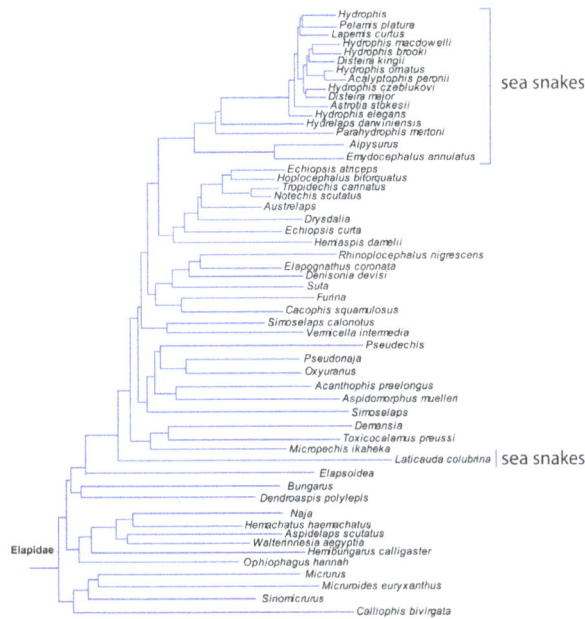

Sea snakes and their relationship to other Elapid snake genera.

Some results indicate that true sea snakes (Hydrophiinae) can be separated into two totally different groups, indicating that sea snakes perhaps have evolved three times from terrestrial elapids. However, this hypothesis is not supported by other recent results and the latest molecular evidence suggests that the true sea snakes (Hydrophiinae) are a very young monophyletic group, perhaps only 8–13 Myr old.

On the genus level at least three completely different taxonomic hypotheses exist, resulting in much confusion in both scientific and non-scientific publications. The most thorough work on sea snakes is by Smith, 1926, whose taxonomy has been used by most authors during the past 85 years. In 1972 McDowell, based on a few specimens from selected species, suggested separating the genus Hydrophis into 3 different subgenera based on morphological data. McDowell also changed some genera using the same arguments, which were not based on a phylogenetic analysis, but more on morphological similarity. Kharin has suggested raising some of McDowells' species groups to genus level; however, Kharin also did not produce any phylogenetic analysis for these groups. Recently, Wells suggested changing the nomenclature in the Australian sea snake species also without any phylogenetic analysis, thus creating new problems for the layman to use valid generic names for the groups. A couple of papers using phylogenetic methods do not support the hypothesis by McDowell. Because of sea snakes' extremely venomous bite it is important to have correct taxonomic assignments and thus correct names because physicians often apply anti-venoms based on the species of snake. As long as no thorough phylogenetic analysis is available for the group, we suggest using Smith's classification, with a few corrections that have been generally accepted.

Identifying sea snakes to species level is very difficult; especially the genus Hydrophis shows great intraspecific variation making it difficult to use only external characters for their identification. The ~70 species recognized here follow Golay et al. and David & Ineich with a few new species added or resurrected. A complete species list can be found in when searching for the two subfamilies, Hydrophiinae and Laticaudinae, respectively.

At a higher taxonomic level sea snakes are the closest relatives of terrestrial elapids, which include some of the most venomous snakes in the world (e.g. brown snakes, taipan, death adder, cobra, krait, mambas). Sea snakes or aquatic elapids and terrestrial elapids are collectively known as proteroglyphous snakes because of the position of the poison-fangs in the frontal part of the upper jaw (maxillary bone).

Feeding and Breeding Biology

Most sea snake species feed on fish that are close to the bottom or sedentary; a few prefer fish eggs (Aipysurus eydouxii, and the genus Emydocephalus; some specimens of A. fuscus also contained fish eggs) and at least Aipysurus laevis, Enhydrina schistosa and Lapemis curtus have been found with crustaceans and mollusks in their stomachs. The Laticauda group feed mostly on eels burrowed in sand at the bottom of the sea and in reef crevices. Using sea snakes to capture undescribed eel species has been shown to be very effective because the eels are extremely secretive in their habits giving no chance to collect them using traditional methods.

All sea snakes produce living young (viviparous) except for the genus Laticauda which is egg laying (oviparous). The annual reproductive cycles are synchronous between males and females and they reproduce every second year with a clutch size that increases with the size of the female.

By-catch and Commercial use of Sea Snakes

Sea snakes are not only of interest because of their poison, but also in connection with the commercial exploitation of reptile skin, organs and meat. Some species are accessible in great numbers (e.g. Laticauda spp., Lapemis spp. and some Hydrophis spp.) and are not protected by CITES (Washington convention).

Since at least 1934 sea snake meat and skins have been used commercially in the Philippines, but also in Australia. In Japan, Taiwan, Thailand and Vietnam sea snakes have been collected commercially. Local protection of sea snakes has been necessary to stop over-exploitation in the Philippines. In Australia commercial sea snake fisheries and by-catch have been investigated during the past 15 years. However, most sea snake fisheries in the Indian Ocean and in the Pacific are not reported in the literature and are beyond control of the local governments.

Species Distribution and Density

Many of the more than 60 species of sea snakes have a broad distribution in both the Indian Ocean and the Pacific Ocean. Species such as Acalyptophis peronii, Aipysurus eydouxii, Astrotia stokesii, Enhydrina schistosa, Lapemis curtus, Laticauda colubrina, L. laticaudata and many Hydrophis species have been collected in both the Asian and the Australian region and are abundant in these areas. Other species are much less well known and some species are only collected from very restricted areas and therefore much more vulnerable for change in the environments. One of the rarest species is Hydrophis parviceps which is known from only two specimens collected in a very limited area in the southern Vietnamese part of the South China Sea. Hydrophis sibauensis is a recently described species which has been collected more than 150 km into rivers of Borneo and is only known from 3 specimens. This is also the case for another recently described species, Hydrophis laboutei which is only known from two specimens collected at Chesterfield Reefs (New Caledonia). Other species with a limited distribution are therefore highly vulnerable to all kinds of environmental changes, including Aipysurus fuscus, A. apraefrontalis and A. foliosquama which have only been collected in the north west part of Australian coral reefs. Recent surveys indicate that there has been a drastic decline of specimens in this area over the last 10–15 years. Hydrophis semperi and Laticauda crockeri, both species only known from Lake Taal in the Philippines and Lake Te-Nggano in Rennell Island, respectively, also represent species that have a very restricted distribution and therefore are highly vulnerable to changes in the environment.

Our own (ARR) investigations in the Gulf of Thailand and part of Borneo indicated that the sea snake fauna in this area has also declined. In the Gulf of Thailand, especially populations of species going up rivers have disappeared (e.g. Hydrophis torquatus and H. klossi), but also in Borneo the number of specimens collected previously in great numbers inside river mouths appears to have declined. Trawl fisheries in the Andaman Sea (e.g. around Phuket, Thailand) also got fewer specimens and the fishermen also mention that the size of sea snakes they got in former times was much larger than the ones they get now. A small survey in Cambodia (ARR) in the year 2000/2001 indicated similar results to those found for Borneo and Thailand.

Conservation: Methods to Ensure the Survival of Sea Snakes

The main threat to sea snakes is the cultural indifference to conservation issues by locals and, as a consequence, their commercial exploitation. Only raising awareness may reduce this kind of threat. Another problem in trying to help the sea snakes to survive is the very limited knowledge on their biology, especially in Asia. It is very important to get much more information on the biology in formulating management plans. For the moment we still miss information on breeding cycles, by-catch and mortality, growth rates, population density, sexual maturity and taxonomy in most areas. The effects of the exploitation or/and by-catch on the sea snakes are almost unknown except from the Philippines and Australia. Some populations may already be in danger of extinction. The only way to have a sustainable yield is by monitoring and controlling by-catch and commercial catch of sea snakes, giving local governments a chance to intervene before a catastrophic collapse of local populations occur. However, to limit exploitation of the most common species sustainable and to protect the endangered ones, we need to have much more biological focus on the group.

One cause for the disappearance of sea snake species from rivers is at least in part due to the great problems with the ongoing pollution in many rivers in both Asia and Australia. Also the information on breeding areas for most sea snake species is unknown, but there is no doubt that also habitat destructions (e.g. mangrove clearings and water crafts) have a negative influence on sea snake populations.

A first step in sea snake conservation is to distinguish the many species from each other, which is not an easy task, and with the global decline of taxonomists it may be more important to focus initially on the entire group. The next step is to get more knowledge of sea snake biology and then to focus on the by-catch and management plan to protect the endangered species and harvest only the more common ones. Concerning the more global problems of pollution and habitat degradation, we have to put more pressure on politicians and hope that they will come up with solutions for the benefit of both humans and sea snakes.

Other Snakes Found in Marine and Brackish Environments

Evolution is continually tinkering with snake populations that live in coastal areas adjacent to brackish and salt water environments. Besides the true sea snakes, many terrestrial and arboreal species have learned to exploit marine resources by foraging in the intertidal zone at low tide or from the branches of mangroves, while some freshwater species have adapted to life in brackish water, sometimes enter the ocean, or live there permanently.

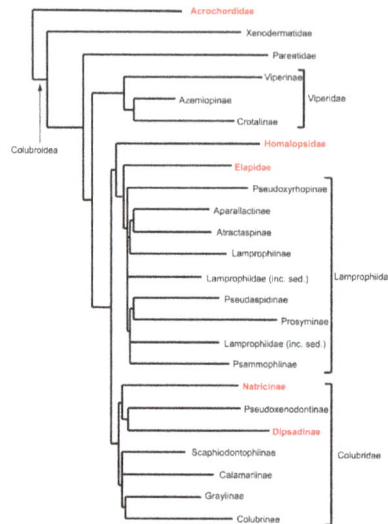

Several families of snakes have independently adapted to saltwater.
Families with species that live in brackish or marine environments are shown in red.

Genus	Total species	Species in brackish or marine water
Acrochordus	3	all (3)
Agkistrodon	4	1
Bitia	1	all (1)
Cantoria	1	all (1)
Cerberus	3	2+
Crotaphopeltis	6	1
Djokoiskandarus	1	all (1)
Enhydris	24	1
Farancia	2	all (2)
Fordonia	1	all (1)
Gerarda	1	all (1)
Grayia	4	1
Helicops	15	2+
Hydrops	3	1
Liophis	39	2+
Myron	3	all (3)
Natrix	4	3+
Nerodia	10	4+
Regina	4	1
Stegonotus	9	2+
Storeria	4	1
Thamnophis	31	2
Tretanorhinus	4	2

The list does not include arboreal and terrestrial species that use mangrove forests or terrestrial species that occasionally enter water, or forage in the intertidal zone. A "+" indicates there are probably more species in this genus that use brackish water.

Non-elapid snakes that use brackish water.

All three living species of the ancient Acrochordidae use a combination of aquatic environments ranging from freshwater to sea water, and have probably been living in coastal ocean habitats for most of their 90.7 million year history. Acrochordids, particularly Acrochordus granulatus have some of the most specialized morphology and physiology for life in saltwater, including the greatest capacity to store oxygen found in any vertebrate. These are low metabolism snakes, feeding and reproducing infrequently and incapable of active swimming for more than a few minutes. Acrochordus arafurae females ambush prey in deep water while males actively search for prey in shallow water. They have a constriction-like behavior for holding prey between body loops during ingestion, and enlarged keels on their scales aid in holding slippery fish. Additionally each scale contains a mechanoreceptor that may be used to locate fish in turbid water.

The Asian and Australasian Homalopsidae, ironically sometimes called the "freshwater snakes," have species that live in coastal habitats such as mangrove forests and salt marshes. Homalopsids inhabit life zones that range from the fossorial Brachyorrhos albus and the semi-terrestrial Enhydris plumbea, to Erpeton tentaculatus that is all but incapable of leaving their freshwater habitats, to species inhabiting coastal marine environments (Bitia hydroides, Cantoria violacea, Myron sp., Fordonia leucobalia, Enhydris bennettii). The most widespread and successful brackish water homalopsids are the bockadams (Cerberus sp) which are distributed from the vicinity of Mumbai, India in the east to Palau, Micronesia in the west, and range southward into the Indonesian Archipelago, New Guinea, and northern Australia. Like some of the Hydrophis and Laticauda, the Lake Buhi Bockadam (Cerberus microlepis) is a freshwater species, derived from its nearby salt water dwelling relative, Cerberus rynchops in the Philippines. While most of the aquatic homalopids are piscivorous, three species, all members of the same clade specialize in feeding on crustaceans (Cantoria violacea, Fordonia leucobalia, and Gerarda prevostiana) in near shore habitats.

A homalopsid snake (Cerberus rynchops).

An assemblage of mangrove homalopsids studied in Singapore was dominated by the piscivorous Asian Bockadam, Cerberus rynchops (73% of total snakes), while the other species were crustacean specialists and less common. Gerard's Mud Snake, Gerarda prevostiana (16% of the total snakes) feeds exclusively on recently-molted crabs; while the Crab-eating Snake, Fordonia leucobalia, (10% of total snakes) specializes in feeding on hard-shell crustaceans. The most uncommon Singapore homalopsid was Cantor's Mud Snake, Cantoria violacea, (2% of total snakes) and it too has a specialized diet, feeding on Alpheus snapping shrimp (Decapoda, Alpheidae). The crustacean-eaters were often observed in association with mud lobster mounds constructed by Thalassina anomala (Decapoda: Thalassinidae). The snakes were nocturnal and active throughout the night. Gerard's Mud Snake increased activity during spring tides, but the other species did not. Three male Crab-eating snakes were tracked using radiotelemetry: they rarely moved and when they did it was for short distances (1.8 to 14.0 m). As might be expected for tropical aquatic snakes, the body temperatures were very stable (26.3–29.0°C) and consistently higher than the microhabitat temperatures. Two of the radio tracked crab-eating Snakes made extensive use of mud lobster mounds.

Snakes of the family Natricidae (or subfamily Natricinae) make up about 30 genera and 210 species. They occur in both hemispheres and in both temperate and tropical environments, but they appear to have originated in Asia, and dispersed into Europe, and North America. Many of these snakes are semi-aquatic and some inhabit brackish water. In North Africa and Europe at least three species use coastal habitats: the Viperine Snake, Natrix maura, the Grass Snake, N. natrix sicula, and the Dice Snake, N. tessellata. North American brackish water natricids use mangrove and salt marsh habitats include the Mangrove Water Snake, Nerodia clarkii, in the southeastern USA, and the Baja Garter Snakes, Thamnophis valida. Both these species are piscivorous and like some freshwater natricids, Nerodia clarkii juveniles lure prey with their tongue. The Salt Marsh Brown Snake, Storeria dekayi limnectes is less aquatic than the previous species, and it feeds on soft bodied invertebrates.

In the Neotropics, the family Dipsadidae (or subfamily Dipsadinae) has more than 92 genera and >700 species, a few of which use brackish and salt water habitats (Helicops, Hydrops, Liophis, and Tretanorhinus) to varying degrees. For the most part these snakes are poorly known and their brackish water habits are known only from anecdotal observations. Similarly, at least two aquatic African snakes, of uncertain lineages (Grayia smythii and Crotaphopeltis hotamboeia) occur in the brackish water of mangrove forests as well as freshwater.

There is another group of snakes that use marine resources that is worthy of note. These species forage in mangroves, salt marshes, and intertidal zone without actually spending much time in the water. The most specialized is perhaps the Burmese Vine Snake, Ahaetulla fronticincta (family Colubridae) which hunts gobies from branches over hanging the water . The Southeast Asian Mangrove Pitviper, Cryptelytrops purpureomaculatus (Viperidae) also hunts from the mangrove branches but its diet is unknown. There are others, for the most part habitat generalists with populations in or adjacent to marine habitats and include Python molurus, Python reticulatus (Pythonidae), Ophiophagus hannah (Elapidae), and Boiga dendrophilia (Colubridae) – all medium to large snakes that forage for food in coastal habitats. Other snakes forage in the intertidal zone: Coluber anthonyi (Colubridae); Crotalus mitchelli and Crotalus muertensis (family Viperidae, subfamily Crotalinae).

Thus, of 34 lineages of snakes (families and subfamilies), four (Acrochordidae, Homalopsidae, Dipsadidae, and Elapidae) contain most of the species adapted for marine environments, while other clades have relatively few, or no, species adapted for the saline water. As more herpetologists investigate mangroves and salt marshes a more complete picture of brackish water snakes will emerge.

Marine Iguanas

Lizards are the most speciose and diverse group of reptiles, with almost 5,500 species (60% of all reptiles). Nevertheless, only a few species have ventured into the oceans. The marine iguanas of the Galapagos Islands are the most aquatic of the lizards, but bask and reproduce on land and are subject to terrestrial predators.

The Galapagos archipelago arose through volcanic activity 960 km off the coast of Ecuador. Lizards and other animals probably reached the island on rafts of vegetation washed down the rivers of western South America. The rafts may have included juveniles, adults, or eggs. The Humboldt and El Niño currents would allow for such rafting to originate from the western coast of South or Central America. Other species, such as rats, have been introduced by humans only about 100 years ago.

Besides seven species of smaller iguanids of the genus Microlophus four large species of iguanas inhabit the Galapagos islands: three species of land iguanas (genus Conolophus) and the related marine iguanas (Amblyrhynchus cristatus). Because the land iguanas are endemic to only a few islands, they are threatened by predation of introduced mammals. However, recent measures to control and protect their habitats seem to have stabilized their populations. In addition to these anthropogenic factors, the archipelago is subject to strong seasonal and annual variations in environmental conditions, so that a combination of factors could turn out to be detrimental.

The marine iguana, Amblyrhynchus cristatus, and the land iguanas of the genus Conolophus are all about 1.2 m in length. Amblyrhynchus inhabits virtually all islands of the archipelago, given its ability to disperse to various islands. As a result, they are less prone to extinction.

Although land and marine iguanas are very different in appearance, they are closely related. In fact, several instances of hybridization between the two genera have been reported. Hence it is likely that they both evolved relatively recently from a land iguana that came from the South American mainland.

Marine iguanas feed exclusively on marine plants. While they spent a considerable time in the water foraging, they have not completely adapted to marine life. For instance, they still have to nest on land and also bask on land to reach their optimal body temperature which rapidly declines in the rather cool ocean water. Nevertheless, considerable selection pressure has resulted in several adaptations to their marine lifestyle such as a flattened tail and limited webbing of all four feet, supporting swimming. Powerful claws help them to hold on to rocks in the heavy sea. Marine iguanas also have reduced the number of heartbeats per minute from about 43 on land to 7 to 9 while diving, as do several other reptiles. Finally, both Conolophus and Amblyrhynchus possess nasal salt glands, similar to those found in other reptiles that have high dietary salt intakes. Interestingly, neither species has the capacity to produce hyperosmotic urine. Thus, the marine iguana has the highest known extracloacal excretion rate of Na, Cl, and K of any reptile.

Marine Mammals

A Marine mammal is a mammal that has adapted to aquatic life and rely's on the ocean to maintain a healthy, livable existence.

These amazing animals can be found living in all of the worlds major oceans from the tropical environments in and around the equator to the northern and southern polar hemispheres in and around the Arctic and Antarctic oceans.

Marine mammals share several characteristics that are common among all mammals such as the need to breathe air, being warm-blooded, having mammary glands which produce milk to feed their young, giving birth to live young (pregnancy/gestation periods) and in some cases having hair.

Here are several common characteristics found in marine mammals:

- Marine mammals breathe air – Although marine mammals live in and around the water they must come to the surface to breathe otherwise they'll drown.

- Marine mammals are warm-blooded – In order to maintain their body heat marine mammals consume large quantities of calories and develop a thick layer of fat or blubber to keep their vital organs from freezing in cold environments.

- Marine mammals give birth to their offspring – Unlike fish and other aquatic animals whales do not lay eggs; instead they carry their children in their womb until they are born.

- Marine mammals produce milk – Marine mammals have mammary glands that produce milk which they use to feed their children. The milk is often full of fat and nutrients to help the child develop.

Marine Mammal Characteristics

While land mammals are similar in many ways marine mammals have adapted to deal with living in environments that are generally surrounded by water rather than being elusively surrounded by land, and way marine mammals live, hunt for food and navigate the world is largely dependent upon their oceanic environment and physical characteristics.

For example marine mammals such as whales, dolphins and porpoises have streamlined bodies designed to reduce water resistance when swimming, they also have specialized lungs and muscles designed to store oxygen, thick blubber or fat to help maintain body heat in cold environments, large veins to help transfer blood to vital organs in a cold water or when in deep waters where compression can become an issue and fins, flukes, and flippers to assist with swimming.

Other species such as polar bears have thick layers of fat and fur to help them maintain their body heat in cold climates, a large body to help keep them warm and disperse their body heat, thick curved claws for gripping the ice and predators, sharp teeth and strong jaws for tearing apart prey and powerful muscles for fighting and attacking.

Seals have thick layers of fat/blubber to help keep them warm, rounded bodies to help disperse their body heat, limbs designed for walking/waddling on land with webbed feet/flippers to help them swim and teeth which they use to grip onto their prey.

Seals, sea lions and walruses have a number of characteristics that are similar among the separate species but also share a few distinct differences as well.

For example walruses tend to have a stockier build than seals and sea lions and they also possess two large teeth that protrude from their mouth, similar to a saber toothed tiger only larger.

Seals also have separate distinct characteristics from sea lions such as having shorter for limbs with long claws, which requires them to waddle on land rather than walk and their strong rare flippers allow them to propel themselves through the water more effectively than sea lions.

While not all marine mammals share the same biological characteristics, (a clear example would be to compare polar bears to whales) they all live near or in the ocean and need the ocean for its supply of food and water in order to survive.

Distinct Types of Marine Mammals

In terms of the various marine mammal species there are several distinct groups of marine mammals; sirenians (manatees and dugongs) cetaceans (whales, dolphins and porpoises), fissipeds, sea otters, pinnipeds (seals, fur seals, sea lions and walruses) and (as considered by some researchers and biologists) polar bears.

Both cetaceans and sirenians live in the ocean while polar bears, pinnipeds and fissipeds are land dwellers, but rely on the ocean to supply them with food and water.

As a whole there are over 125-recorded species of marine mammal inhibiting the ocean and native aquatic environments of the world.

Here is a list of the of individual sub groups within the marine mammal family:

- Cetaceans (whales, dolphins and porpoises)
- Fissipeds
- Pinnipeds (seals, fur seals, sea lions and walruses)
- Ursidae (polar bears)
- Sea otters
- Sirenians (manatees and dugongs)

Although these animals can be found throughout the world the highest concentration of marine mammals (about 40%) are found at or around 40° both north and south of the equator.

Marine Mammals and the Ecosystem

When it comes to survival and the food chain marine mammals play a large role in maintaining a healthy marine ecosystem, however since around 20% – 30% are considered endangered and/or threatened it raises some concerns on the impact of aquatic life and the impact humans have played in contributing to its condition.

The elimination of one species could significant decrease the chances of survival for another species, especially among species that rely on their prey for survival.

If their prey were to become extinct than the predators also face the possibility of extinction.

On the other hand if a predator were to become extinct they their prey could overpopulate and quickly consume their food supply leading to future starvation.

Although these are extreme examples they are important to point out as it is a possibility.

In order to better protect these animals all marine mammals in the United States are protected by the Marine Mammal Protection Act created in 1972.

In addition to the Marine Mammal Protection Act other countries have also created laws designed to protect these animals and organizations have been created to bring awareness about the importance of protecting the endangered marine mammals as well as implement programs that are designed to aid in the recovery of certain species.

Marine Invertebrates

Animals that lack backbones are known as invertebrates. Over 98% of species on Earth are invertebrates that rely on other strategies than a backbone for support such as hydrostatic pressure, exoskeletons, shells, and in some, even glass spicules. Some invertebrate phyla have only one species, while others like Arthropoda include more than 83% of all described animal species with over a million species. The most common marine invertebrates are sponges, cnidarians, marine worms, lophophorates, mollusks, arthropods, echinoderms and the hemichordates.

Sponges

There are between 9,000 and 15,000 species of sponges classified under the Phylum Porifera. Sponges are relatively simple animals that originated with the first animal life in the Precambrian times.

The anatomy of a typical sponge is organized so that flagella inside the sponge pull water into small holes (ostia) in the body and expel waste through larger holes (oscula). Sponge species have a variety of body plans that provide structure, including support by organic fibers (Class Demospongiae - 90% of sponge species), calcareous spicules (Class Calcarea ~400 species), and siliceous spicules (Class Hexactinellida) or combinations of these.

The body plan of a sponge has adapted to filter small food particles from the passing water allowing them to reside in most habitats, including polar shelves and submarine caverns that often contain very few nutrients.

Sponge, HondurasLike other animals, sponges were found to also grow extremely slowly in cold waters such as those of the Antarctic. Age estimates based on growth rates of one glass sponge (Scolymastra joubini) in the Ross Sea were between 15,000 and 23,000 years, which means that specimen appears to be the longest-lived animal on earth yet recorded. Sponges are often studied by scientists to find clues about the first life forms on Earth with more than one cell.

Sponges are hermaphroditic and are able to reproduce both sexually and asexually. Most sponges usually reproduce by sexual reproduction, where sperm cells (spermatocytes) develop from choanocytes (collar cells) and eggs develop from oocytes. When environmental conditions are favorable, spermatocytes are ejected in out-going currents and the eggs, once fertilized inside the sponge in some sponges, develop into flagellated larva that swim about as plankton until they find a suitable place to settle and grow into adults. Asexual reproduction occurs when favorable environmental conditions deteriorate and includes both regeneration (regenerating from fragments), budding (groups of cells differentiate into small sponges that are then released externally or expelled through the central canal (oscula)), or the formation of gemmules ("survival pods" of unspecialized cells that remain dormant until conditions improve and then either form completely new sponges or re-colonize the skeletons of their parents).

Sponges are eaten by chitons, snails, nudibranchs, turtles, fish, and insects. They provide a home to sea anemones, polychaetes, octopuses, copepods, zoanthids, shrimps, brittle stars, amphipods, barnacles, and fish. There are numerous symbiotic relationships between animals and sponges.

Sponges that are composed of organic fibers (demosponges) have been used by humans for thousands of years for cleaning and other purposes. Sponge diving has declined significantly due to overfishing and most sponges these days are now synthetic.

Cnidarians

The Phylum Cnidaria consists of about 10,000 species of "simple" animals found only in marine habitats and includes Class Anthozoa (corals and sea anemones), Class Hydrozoa (hydrozoans), Subphylum Medusozoa: Class Cubozoa (box jellyfish), Class Scyphozoa (jellyfish), and Class Staurozoa which contains Order Stauromedusae (stalked jellyfish). Phylum Cnidaria may also contain Family Polypodiidae and Family Tetraplatidae. Species in cnidaria have special stinging cells called cnidocytes. Cnidarians evolved during the Precambrian era and are some of the earliest multicellualr life forms known.

Most cnidarians have a very basic body plan which includes a digestive cavity with one opening. This opening functions as both the mouth and anus for the organism. The only true organs in cnidarians are the gonads. Most cnidarians are symmetrical, an observation referred to as "radial symmetry." Cnidarians also have an ectoderm (tissue that covers the outer body surfaces) and an endoderm (inner layer of cells forming the gastrointestinal and respiratory tracts, and inner organs). The ectoderm is connected to the endoderm by a gel-like substance known as the mesoglea. Cnidarians use a nerve net and very basic receptors for impulses to move. Oxygen is taken in directly from the water through the tissues.

Organisms in cnidaria capture and kill their prey using cnidocysts, or stinging cells, around their mouth which send out stinging barbs which immobilize their prey and help protect against predators. Once prey is captured the tentacles move it into the central gastrovascular cavity where it is digested. Waste is then expelled back through the mouth.

The four classes of cnidarians are the Anthozoa, the Hydrozoa, the Scyphozoa, and the Cubozoa. Anemones, corals, and sea fans are in class Anthozoa, which was the first to diverge throughout evolution. Portuguese man-o-wars and obelia are examples of animals in Hydrozoa, jellyfish are in class Scyphozoa, and box jellies are in class Cubozoa.

Cnidarian species have a variety of life-cycles. Some alternate between being free-swimming medusae and asexual polyps depending on their environment. In some groups like Anthozoa, organisms never make it to the free-swimming medusae stage, but instead live their whole lives as a non-moving polyp. Organisms in the groups Scyphozoa and Cubozoa spend most of their lives in the medusal stage. Medusae can measure anywhere from a few millimeters to 30 meters long including the tentacles. Some, like the Siphonophores, are individuals but can live in colonies and appear as one organism.

Marine Worms

Marine worms can be placed into more than ten different phyla and come in a variety of colors, shapes, and sizes. Marine worms are often confused with other animals with thin and long bodies. Most marine worms are grouped into the Annelids, a group that includes the Polychaetes (bristle worms), Oligochaetes, Hirudinae, and the Eunice aphroditois. Polychaetes are most often found near the shoreline and swim or crawl using a pair of legs found on each segment of their body. The Oligochaetes, which include earthworms, are mainly found on land and the subclass Hirudinae include leeches that usually live in freshwater environments. Some marine worm species, such as the bearded fire worm, can deliver a nasty burning sting to humans when handled.

The body structure of an annelid consists of a front end with a prostomium, also referred to as a significantly defined head. Most annelids have two pairs of eyes, three antennae, a pharynx or proboscis used to eat food and tentacle-like cirri for probing the surrounding area. An example of the biodiversity of worm species is the Sipunculid also known as the peanut worm. This worm digs itself into a hole underneath rocks, eats organic material, has no segments and looks like a peanut when it pulls its proboscis into itself.

In general, marine worms live underneath rocks near the edge of the ocean, in algae, or anywhere there is mud or sand. Species of marine worms can be ringed, segmented, or flat and include tube-digging worms, burrow-dwelling worms, ribbon worms, and peanut worms.

Some common annelids include the tube-making Galeoloaria, the stinging fireworm, the short scale worm, and the huge Eunice aphroditois. Tube worms actually make a tube with a hard shell and retreat into the shell when threatened. The Christmas tree worm has many brightly colored feather-like tentacles shaped somewhat like a Christmas tree that is used to filter food from the water.

Lophophorates are characterized by a special feeding organ called a lophophore which is an extension of the body wall into a tentacled structure that surrounds the mouth and is either U-shaped or circular. The lophophore is used to trap floating food particles in passing currents (called suspension feeding). Tentacles surrounding the mouth are usually hollow and the mouth is usually located inside the lophophore. The anus is on the same side of the body but on the outside of the lophophore. Lophophorates include the phyla Phoronida, Bryozoa (Ectoprocta), and Brachiopoda and are related to the Mollusca and Annelida phyla. Many lophophorates have tubes, shells, or exoskeletons for protection. They are usually sessile (non-moving), benthic (sea floor dwellers), and live in salt water, although there are a few freshwater lophophorates in the Phylum Bryozoa.

Phylum Bryozoa contains Class Gymnolaemata (marine bryozoans) and Class Phylactolaemata (freshwater bryozoans ~50 species) and are tiny though visible colonial animals that look a bit like tiny coral colonies that also build skeletons of calcium carbonate (although some species lack calcification and instead are mucilaginous (made of slime)). Members of the Phylum Bryozoa are known as "moss animals" or "sea mats" and they generally prefer warm, tropical waters, but are known to occur worldwide. There are about 8,000 living species, with many times that known from the fossil record. Fossil bryozoans are common throughout the world in sedimentary rocks representing shallow marine habitats, especially in rocks of Paleozoic age.

Bryozoans are usually found on hard substrates such as rocks, shells, wood, blades of kelp, and ships, which can become heavily encrusted with bryozoans. Some bryozoan colonies also form colonies directly on marine sediments. Bryozoans have been found at depths of 8,200 m (27,000 ft) though most inhabit shallower warmer waters. Most bryozoans are though a few are able to creep about and some non-colonial bryozoans live and move about in the spaces between sand grains. One species appears to make its living while floating in the Southern Ocean.

Almost all bryozoans are colony-forming animals often with millions of individuals in each colony. The colonies range from millimeters to meters in size, but the individuals that make up the colonies (called zooids) are tiny, usually less than a millimeter long. In each colony, different individuals or zooids assume different functions. Some gather food for the colony (autozooids) while others have specialized for different functions (heterozooids) such as kenozooids which provide structural support and vibracula which have long whiplike structures that they use to clear debris away from the surface of the colony. There is only a single known solitary species, Monobryozoon ambulans, which does not form colonies.

Bryozoan skeletons grow in a variety of shapes and patterns: mound-shaped, lacy fans, branching twigs, and even corkscrew-shaped. Their skeletons have numerous tiny openings, each of which is the home of a zooid. They also have a coelomate body with a looped alimentary canal or gut, opening at the mouth and terminating at the anus. They feed with a specialized, ciliated structure called a lophophore, which is a crown of tentacles surrounding the mouth. Their diet consists of small microorganisms, including diatoms and other unicellular algae. In turn, bryozoans are preyed on by grazing organisms such as sea urchins and fish. Bryozoans do not have any defined respiratory, or circulatory systems due to their small size. However, they do have a simple nervous system and a hydrostatic skeletal system. Several studies have been undertaken on the crystallography of bryozoan skeletons, revealing a complex fabric suite of oriented calcite or aragonite crystallites within an organic matrix.

The tentacles of the bryozoans are ciliated, and the beating of the cilia creates a powerful current of water, which drives water together with entrained food particles (mainly phytoplankton) towards the mouth.

Because of their small size, bryozoans have no need of a blood system. Gaseous exchange occurs across the entire surface of the body, but particularly through the tentacles of the lophophore.

Bryozoans can reproduce both sexually and asexually. All bryozoans, as far as is known, are hermaphroditic (meaning they are both male and female). Asexual reproduction occurs by budding off new zooids as the colony grows, and is the main way by which a colony expands in size. If a piece of a bryozoan colony breaks off, the piece can continue to grow and will form a new colony. A colony formed this way is composed entirely of clones genetically identical individuals) of the first animal, which is called the ancestrula.

One species of bryozoan, Bugula neritina, is of current interest as a source of cytotoxic chemicals, bryostatins, under clinical investigation as anti-cancer agents.

The closest relations of the Bryozoa appear to be the brachiopods.

Mollusks

Animals classified under the phylum Mollusca are extremely diverse in form, but all have a fairly simple body plan. Familiar mollusks include oysters, chitons, clams, snails, slugs, octopus, and squid. Most mollusks have a soft body and a hard or "calcareous" shell. Many mollusca use mucous and cilia to eat, move, and reproduce. There are more than 110,000 species in phylum Mollusca, more than every other phylum except Arthropoda. With a few exceptions, all living species of mollusks are categorized under Gastropoda or Bivalvia. Another important class is Cephalopoda. Some scientists have determined that there is more biomass from marine mollusks than any other animal on earth.

Mollusks reproduce through external fertilization where the eggs and sperm are released into the water. In some more complex mollusks, fertilization can take place internally after long courtship rituals and mollusk dances. Many of the more sophisticated snails are hermaphroditic. Some go through phases where they alternate gender; others are both female and male at the same time.

Almost all mollusks living in freshwater are gastropods although a few bivalves can be found in brackish water. Some species of mollusks have adapted to living on land but can only live in humid environments. Terrestrial mollusks must have the ability to regulate their temperature, breathe air, make larger eggs, and maintain moisture levels by conserving water. Snails that live in the littoral zone of the ocean often show similar adaptations as terrestrial or land living snails. Snails in the class Pulmonata have adapted to living on land so well that they can be found at high altitudes. Other snails in Pulmonata that once could breathe air have gone back to living in the water.

Mollusks can be found in all habitats of the ocean. Some bivalves like the Protobranchiates, are even found in waters 9,000 m or 29,500 ft deep. The more advanced cephalopods could be viewed as the most sophisticated invertebrates. Animals like squid, cuttlefish, and octopuses have relatively "huge" brains and move about using their arms, fins, and siphons (in a manner similar to jet propulsion).

Arthropods

Arthropoda is the largest phylum in the taxonomic system and is composed of insects, crustaceans, and arachnids. Nearly 4/5 of all living animals are arthropods.

This ancient phylum dates back to the earliest days of the Cambrian period

Arthropods are characterized by a segmented body plan with a head, abdomen, and thorax and legs or appendages on every segment with a rigid exoskeleton made out of chitin. Arthropods use their appendages to feed, as sensory mechanisms, and for locomotion. Aquatic arthropods use gills for respiration.

Although spiders are possibly the most familiar arthropod, lobsters, crabs, barnacles, and shrimp in the class Crustacea are also in this phylum.

Arthropods are most closely related to the Annelida, or segmented worms. The five main subgroups of the phylum are the Trilobita, Myriapoda, Chelicerata, Crustacea, and the Hexapoda.

Echinoderms

The Echinoderms lack a head and have five-point radial symmetry. These fascinating animals live only in marine environments. They have an endoskeleton made out of calcareous plates, which is often protected by spines. The plates that make up the endoskeleton often support the spines and enclose the coelom, an anatomical feature used for movement, respiration, collecting food, and as a sensory mechanism. The coelom also houses the reproductive organs and alimentary canal.

- Echinoderms can be found in all oceans in all zones with approximately 6,000 described species.

- The two main subphylums in phylum Echinodermata are Eleutherozoa and Pelmatozoa.

- Subphylum Eleutherozoa conatins the superclasses Asterozoa and Cryptosyringida.

- Superclass Asterozoa contains the sea stars/starfishes in Class Asteroidea and the extinct Class Somasteroidea.

- Superclass Cryptosyringida contains Class Echinoidea (heart urchins, sand dollars, and sea urchins), Class Holothuroidea (sea cucumbers), and Class Ophiuroidea (basket stars, brittlestars, and snake stars).

- Subphylum Pelmatozoa contains the Class Crinoidea (feather stars and sea lillies).

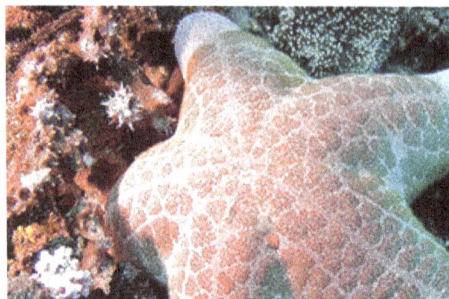

Mature echinoderms have five points that face outward from the center of the body with a mouth underneath and the anus on top. There are exceptions to this plan however; some echinoderms lack an anus and others, like the crinoids, have both the mouth and the anus on the same side of the body. Scientists refer to the side of the body with the mouth as the oral side and the side with the anus as the aboral side. Crinoids, ophiuroids, and holothuroids have tube feet to help collect food particles floating towards their body. Other types of echinoderms like asteroids are carnivorous and will surround or throw their stomach over their prey. Some echinoids even have teeth used to chew and dismantle plants and small animals.

Most echinoderms reproduce sexually producing larvae that feed on phytoplankton until they reach maturity. Some species of echinoderms develop their offspring in embryonic sacs located on the outside of their bodies.

Sea urchin, Lembei Strait, Suluwesi, Indonesia 2005Echinoderms have fascinating water-vascular systems that likely originated from some sort of respiratory system that evolved to include food gathering and movement. They accomplish these tasks through the use of their numerous hollow tube feet that resemble tentacles. There are two rows of tube feet on the outside of the body that fill with seawater so that when the animal expands or contracts, water is drawn into the feet. Once filled, the feet extend outward allowing the animal to walk. Suckers located at the tips of the tube feet are often used to grab prey or to hold onto solid objects when the echinoderm wants to remain attached to something.

The most familiar echinoderm known to humans is probably the sea star, categorized in the super-classes Asterozoa and Cryptosyringida. There are two classes of sea stars which include Asteroidea and Ophiuroidea. True sea stars and sun stars in are in Class Asteroidea while brittle stars and basket stars are in Class Ophiuroidea.

Echinoderms in the class Asteroidea have arms that are smoothly connected to the body; echinoderms in Ophiuroidea have arms that shoot out from a disk-like center. Both are able to regenerate their limbs when one is broken off. In some cases, a lost limb can generate a whole new sea star. The small bumps on top of the sea star are referred to as dermal branchiae and are used to absorb oxygen from the water for respiration. Pedicellaria are small appendages used to keep foreign bodies off of the sea star. The madreporite is a hard opening on the aboral side of the sea star used to regulate and filter sea water.

Sea stars also have an eye-like structure at the end of each arm, called the eyespot, used to detect light.

Hemichordates

Hemichordates are a relatively small phylum. These creatures are extremely important to the study of the evolution of vertebrates. They are characterized by a body divided into three main areas: the preoral lobe, the collar, and the trunk. Hemichordates are partial chordates and are closely related to the first chordates. According to DNA analysis, hemichordates are closely related to echinoderms, which is also apparent during observations of hemichordate and echinoderm larval stages. Hemichordates have gill slits, a structure that resembles a notochord but is called the stomochord, a dorsal nerve cord, and a reduced ventral nerve cord.

There are three classes of hemichordates which include Enteropneusta, Pterobranchia, and Graptolithina. The most well-known class is the Enteropneusta or "acorn worms". Acorn worms have gill slits, burrow into the sediment, and likely feed on dirt and detritus. They can reach up to 2.5 m or 8 ft in length but most are actually quite small. In the Pterobranchia class, there are only a few species notably different from the acorn worms. Pterobranchs live in colonies connected by stem-like stolons. Each tiny individual is referred to as a zooid and has one gill slit. The Graptolithina are most well-known in the fossil record showing up in the Ordovician and Silurian times.

Aquatic Plants

Aquatic plants, or aquatic crops, refer to plants and flowers that grow and thrive in an aquatic environment, whether freshwater or saltwater.

Aquatic plants, also termed as hydrophytes or aquatic macrophytes, live within watery environments. In the ecosystem, aquatic plants serve as food and habitat for animals living in the sea and prevent shorelines, ponds and lakes from eroding by providing soil stability.

Characteristics common to aquatic plants:

1. Most aquatic plants do not need cuticles or have thin cuticles as cuticles prevent loss of water.

2. Aquatic plants keep their stomata always open for they do not need to retain water.

3. On each side of their leaves are a number of stomata.

4. Aquatic plants have less rigid structure since water pressure supports them.

5. Since they need to float, leaves on the surface of plants are flat.

6. The presence of air sacs enables them to float.

7. Their roots are smaller so water can spread freely and directly into the leaves.

8. Their roots are light and feathery since they do not need to prop up the plants.

9. Roots are specialized to take in oxygen.

Adaptation of aquatic plants is evident by their structure: deeply dissected and waxy leaves, specialized pollination mechanism and variation in growth pattern. These are the types of plants based on adaptation:

1. Totally submerged plants – Are considered true water plants or hydrophytes. Example: Water starwort submerged in a marsh pond.

2. Floating plants – Are rooted in floating water (example: water lily) or not rooted in the sediment just on the surface (example: duckweed).

3. Swamp plants – Are emergent plants with their lower part submerged. (Example: reed mace).

Aquatic plants are generally divided into four groups for management purposes. These groups are:

*   Algae

*   Floating Plants

*   Submerged Plants

*   Emergent Plants

Many ponds have more than one type of aquatic plant, and care must be taken to identify all the aquatic plants inhabiting the pond. Some pond plants may be beneficial to local or migratory wildlife, and therefore, may want to be encouraged or at least not eliminated.

Algae and other Plankton

Algae are very primitive plants. Some algae are microscopic (planktonic algae). Others are thin and stringy or hair-like (filamentous algae). While still others are large and resemble higher plants but without true roots (chara).

Floating Plants

True floating plants are not attached to the bottom. Floating plants come in sizes from very small (duckweed) to over a foot in diameter (water hyacinth). Most, but not all, have roots that hang in the water from the floating green portions.

Submerged Plants

Submerged plants are rooted plants with most of their vegetative mass below the water surface, although some portions may stick above the water. One discerning characteristic of submerged plants is their flaccid or soft stems, which is why they do not usually rise above the water's surface.

Emergent Plants

Emergent plants are rooted plants often along the shoreline that stand above the surface of the water (cattails). The stems of emergent plants are somewhat stiff or firm.

References

- Elphick, Jonathan (2016). Birds: A Complete Guide to their Biology and Behavior. Buffalo, New York: Firefly Books. p. 80. ISBN 978-1-77085-762-9.

- Marine-life-definition-and-examples-2291890: thoughtco.com, Retrieved 10 June 2018

- Lequette, B.; Verheyden, C.; Jowentin, P. (1989). "Olfaction in Subantarctic seabirds: Its phylogenetic and ecological significance" (PDF). The Condor. 91 (3): 732–735. doi:10.2307/1368131

- Marine-vertebrates, oceans: marinebio.org, Retrieved 31 March 2018

- Jha, Alok (2009-02-02). "Google Earth adds insight into Earth's oceans | Environment | guardian.co.uk". London: Guardian. Retrieved 2009-03-23.

- Types-of-seabirds-387306: thespruce.com, Retrieved 17 March 2018

- Goedert, J. (1989). "Giant Late Eocene Marine Birds (Pelecaniformes: Pelagornithidae) from Northwestern Oregon". Journal of Paleontology. 63 (6): 939–944. JSTOR 1305659.

- What-is-a-marine-mammal: whalefacts.org, Retrieved 25 May 2018

- Harrison, C. S. (1990) Seabirds of Hawaii, Natural History and Conservation Ithica:Cornell University Press, ISBN 0-8014-2449-6

- Aquatic-plants-and-flowers: proflowers.com, Retrieved 24 June 2018

- Pennycuick, C. J. (1982). "The flight of petrels and albatrosses (Procellariiformes), observed in South Georgia and its vicinity". Philosophical Transactions of the Royal Society B. 300 (1098): 75–106. Bibcode:1982R-SPTB.300...75P. doi:10.1098/rstb.1982.0158

- Plant-identification: aquaplant.tamu.edu, Retrieved 14 April 2018

- BirdLife International (BLI) (2012). "Onychoprion fuscatus". IUCN Red List of Threatened Species. Version 2012.1. International Union for Conservation of Nature. Retrieved 4 April 2015

Chapter 3

Marine Habitats

Marine habitats are either coastal or open ocean habitats. The aim of this chapter is to explore the varied types of marine habitats on Earth. Some of the topics elaborated in this chapter cover the diverse aspects of coastal habitats and pelagic zone for a comprehensive understanding of marine habitats.

The marine environment supplies many kinds of habitats that support marine life.

The marine environment supplies many kinds of habitats that support life. Marine life partially depends on the saltwater that is in the sea. A habitat is an ecological or environmental area inhabited by one or more living species.

Marine Habitats: Coral reefs provide marine habitats for tube sponges, which in turn become marine habitats for fishes.

Marine habitats can be divided into coastal and open ocean habitats. Coastal habitats are found in the area that extends from as far as the tide comes in on the shoreline, out to the edge of the continental shelf. Most marine life is found in coastal habitats, even though the shelf area occupies only seven percent of the total ocean area. Open ocean habitats are found in the deep ocean beyond the edge of the continental shelf.

Alternatively, marine habitats can be divided into pelagic and demersal habitats. Pelagic habitats are found near the surface or in the open water column, away from the bottom of the ocean. Demersal habitats are near or on the bottom of the ocean. An organism living in a pelagic habitat is said to be a pelagic organism, as in pelagic fish. Similarly, an organism living in a demersal habitat is said to be a demersal organism, as in demersal fish. Pelagic habitats are intrinsically shifting and ephemeral, depending on what ocean currents are doing.

Marine habitats can be modified by their inhabitants. Some marine organisms, like corals, kelp, mangroves and seagrasses, are ecosystem engineers, which reshape the marine environment to the point where they create habitats for other organisms.

Marine habitats include coastal zones, intertidal zones, sandy shores, rocky shores, mudflats, swamps and salt marshes, estuaries, kelp forests, seagrasses, and coral reefs. In addition, in the open ocean there are surface waters, deep sea and sea floor.

Intertidal zones (those areas close to shore) are constantly being exposed and covered by the ocean's tides. A huge array of life lives within this zone.

Sandy shores, also called beaches, are coastal shorelines where sand accumulates. Waves and currents shift the sand, continually building and eroding the shoreline. Long shore currents flow parallel to the beaches, making waves break obliquely on the sand. These currents transport large amounts of sand along coasts, forming spits, barrier islands and tombolos. Long shore currents also commonly create offshore bars, which give beaches some stability by reducing erosion.

The relative solidity of rocky shores seems to give them a permanence compared to the shifting nature of sandy shores. This apparent stability is not real over even quite short geological time scales, but it is real enough over the short life of an organism. In contrast to sandy shores, plants and animals can anchor themselves to the rocks.

Mudflats are coastal wetlands that form when mud is deposited by tides or rivers. They are found in sheltered areas such as bays, bayous, lagoons, and estuaries. Mudflats may be viewed geologically as exposed layers of bay mud, resulting from deposition of estuarine silts, clays and marine animal detritus. Most of the sediment within a mudflat is within the intertidal zone, and thus the flat is submerged and exposed approximately twice daily.

Mangrove swamps and salt marshes form important coastal habitats in topical and temperate areas respectively. An estuary is a partly enclosed coastal body of water with one or more rivers or streams flowing into it, and with a free connection to the open sea.

Kelp forests are underwater areas with a high density of kelp. They are recognized as one of the most productive and dynamic ecosystems on Earth. Smaller areas of anchored kelp are called kelp beds. Kelp forests occur worldwide throughout temperate and polar coastal oceans.

Seagrasses are flowering plants from one of four plant families which grow in marine environments. They are called seagrasses because the leaves are long and narrow and are very often green, and because the plants often grow in large meadows, which look like grassland.

Reefs comprise some of the densest and most diverse habitats in the world. The best-known types of reefs are tropical coral reefs, which exist in most tropical waters; however, reefs can also exist in cold water. Reefs are built up by corals and other calcium-depositing animals, usually on top of a rocky outcrop on the ocean floor. Reefs can also grow on other surfaces; this has made it possible to create artificial reefs. Coral reefs also support a huge community of life, including the corals themselves, their symbiotic zooxanthellae, tropical fish, and many other organisms.

Coastal

Coastal Marine Habitats are Coral reefs, mangroves forests, and seagrass beds that supports an incredible diversity and abundance of ocean life including fishes, invertebrates, mammals, and seabirds. They protect coasts from extreme events like storm surges and tidal waves. Coasts are places where the land meets the sea.

Ocean

The ocean is a huge area of salt water that covers almost three quarters the earth's surface.

Coastal Water

Coastal water is the seawater around the coast. It is usually more sheltered and shallower than the open ocean.

Beach

Beaches are very often very sandy areas. Beaches also include areas of rocks, dunes and crashing waves

Sand Dunes

Sand dunes are formed by the wind blowing off the ocean and dropping sand high along the beach. Storms and people easily damage sand dunes.

Rock Pools and Rock Platforms

Rock pools and platforms are part of the rocky shore environment. Animals and plants in these habitats survive being flooded in high tide and drying out at low tide.

Estuary

Estuaries are where the rivers and streams meet the sea. They provide important shelter for lots of young animals and plants.

Mangroves

Mangroves are groups of shrubs and trees that grow on the mud flats in coastal areas. These areas are often underwater at high tide.

Intertidal zone

The intertidal zone, also known as the littoral zone, in marine aquatic environments is the area of the foreshore and seabed that is exposed to the air at low tide and submerged at high tide, i.e. the area between tide marks.

In the intertidal zone the most common organisms are small and most are relatively uncomplicated organisms. This is for a variety of reasons; firstly the supply of water which marine organisms require to survive is intermittent. Secondly, the wave action around the shore can wash away or dislodge poorly suited or adapted organisms. Thirdly, because of the intertidal zone's high exposure to the sun the temperature range can be extreme from very hot to near freezing in frigid climates (with cold seas). Lastly, the salinity is much higher in the intertidal zone because salt water trapped in rock pools evaporates leaving behind salt deposits. These four factors make the intertidal zone an extreme environment in which to live.

A typical rocky shore can be divided into a spray zone (also known as the Supratidal Zone, which is above the spring high-tide line and is covered by water only during storms, and an intertidal zone, which lies between the high and low tidal extremes. Along most shores, the intertidal zone can be clearly separated into the following subzones: high tide zone, middle tide zone, and low tide zone.

High Tide Zone (Upper Mid-littoral)

The high tide zone is flooded during high tide only, and is a highly saline environment. The abundancy of water is not high enough to sustain large amounts of vegetation, although some do survive in the high tide zone. The predominant organisms in this subregion are anemones, barnacles, brittle stars, chitons, crabs, green algae, isopods, limpets, mussels, sea stars, snails, whelks and some marine vegetation. The high tide zone can also contain rock pools inhabited by small fish and larger seaweeds. Another organism found here is the hermit crab, which because of its portable home in the form of a shell does extremely well as it is sheltered from the high temperature range to an extent and can also carry water with it in its shell. Consequently there is generally a higher population of hermit crabs to common crabs in the high tide zone. Life is much more abundant here than in the spray.

Middle Tide Zone (Lower Mid-littoral)

The middle tide zone is submerged and flooded for approximately equal periods of time per tide cycle. Consequently temperatures are less extreme due to shorter direct exposure to the sun, and therefore salinity is only marginally higher than ocean levels. However wave action is generally more extreme than the high tide and spray zones. The middle tide zone also has much higher population of marine vegetation, specifically seaweeds. Organisms are also more complex and often larger in size than those found in the high tide and splash zones. Organisms in this area include anemones, barnacles, chitons, crabs, green algae, isopods, limpets, mussels, sea lettuce, sea palms, sea stars, snails, sponges, and whelks. Again rock pools can also provide a habitat for small fish, shrimps, krill, sea urchins and zooplankton. Apart from being more populated, life in the middle tide zone is more diversified than the high tide and splash zones.

Low Tide Zone (Lower Littoral)

This subregion is mostly submerged - it is only exposed at the point of low tide and for a longer period of time during extremely low tides. This area is teeming with life; the most notable difference with this subregion to the other three is that there is much more marine vegetation, especially seaweeds. There is also a great biodiversity. Organisms in this zone generally are not well adapted to periods of dryness and temperature extremes. Some of the organisms in this area are abalone, anemones, brown seaweed, chitons, crabs, green algae, hydroids, isopods, limpets, mussels, nudibranchs, sculpin, sea cucumber, sea lettuce, sea palms, sea stars, sea urchins, shrimp, snails, sponges, surf grass, tube worms, and whelks. Creatures in this area can grow to larger sizes because there is more energy in the localized ecosystem and because marine vegetation can grow to much greater sizes than in the other three intertidal sub regions due to the better water coverage: the water is shallow enough to allow plenty of light to reach the vegetation to allow substantial photosynthetic activity, and the salinity is at almost normal levels. This area is also protected from large predators such as large fish because of the wave action and the water still being relatively shallow.

Mangrove

Mangrove plants are halophytic (salt-tolerant) plant species, of which there are more than 12 families and 80 species worldwide. A collection of mangrove trees in an area makes up a mangrove habitat, mangrove swamp or mangrove forest.

Mangrove trees have a tangle of roots which are often exposed above water, leading to the nickname "walking trees."

Their unusual, dangling roots make mangroves look like trees on stilts. The term mangrove can be used to refer to certain species of trees or shrubs, a habitat or a swamp.

Mangrove Swamps

Mangrove trees grow in intertidal or estuarine areas. They are found in warmer areas between the latitudes of 32 degrees north and 38 degrees south, as they need to live in areas where the average annual temperature is above 66 degrees Fahrenheit.

It is thought that mangroves were originally found in southeast Asia, but have been disbursed around the world and are now found along the tropical and subtropical coasts of Africa, Australia, Asia and North and South America. In the U.S., mangroves are commonly found in Florida.

Mangrove Adaptations

The roots of mangrove plants are adapted to filter salt water, and their leaves can excrete salt, allowing them to survive where other land plants cannot. Leaves that fall off the trees provide food for inhabitants and breakdown to provide nutrients to the habitat.

Shoreline Protection

- Mangroves protect shorelines from erosion

Schooling tarpon.

Mangroves protect shorelines from damaging storm and hurricane winds, waves, and floods. Mangroves also help prevent erosion by stabilizing sediments with their tangled root systems. They maintain water quality and clarity, filtering pollutants and trapping sediments originating from land.

Nursery

- Mangroves serve as valuable nursery areas for fish and invertebrates

Serving as valuable nursery areas for shrimp, crustaceans, mollusks, and fishes, mangroves are a critical component of Florida's commercial and recreational fishing industries. These habitats provide a rich source of food while also offering refuge from predation. Snook (*Centropomus undecimalis*), gray snapper (*Lutjanus griseus*), tarpon (*Megalops atlanticus*), jack (*Caranx spp.*), sheepshead (*Archosargus probatocephalus*), and red drum (*Sciaenops ocellatus*) all feed in the mangroves. Florida's fisheries would suffer a dramatic decline without access to healthy mangrove habitats.

Threatened and Endangered Species

- Mangroves Support Threatened and Endangered Species

In addition to commercially important species, mangroves also support a number of threatened and endangered species.

Threatened species include:

American alligator
(Alligator mississippiensis).

Green sea turtle
(Chelonia mydas).

Loggerhead sea turtle
(Caretta caretta).

Endangered species include:

American crocodile
(Crocodylus acutus).

Hawksbill sea turtle
(Eretmochelys imbricata).

Eastern indigo snake
(Drymarchon corais).

Atlantic saltmarsh snake
(Nerodia clarkii taeniata

Southern bald eagle (Haliaeetus leucocephalus leucocephalus).

Peregrine falcon
(Falco columbarius).

Brown pelican
(Pelicanus occidentalis

West Indian manatee
(Trichechus manatus).

- Atlantic ridley sea turtle (*Lepidochelys kempii*)
- Barbados yellow warbler (*Dendroica petechia petechia*)
- key deer (*Odocoileus virginianus clavium*)

These species utilize mangrove systems during at least some portion of their life histories, while others reside their entire life spans, feeding and nesting within the mangroves.

Renewable Resource

Honey bee.

Mangroves as a Renewable Resource

In other parts of the world, people have utilized mangrove trees as a renewable resource. Harvested for durable, water-resistant wood, mangroves have been used in building houses, boats, pilings, and furniture. The wood of the black mangrove and buttonwood trees has also been utilized in the production of charcoal. Tannins and other dyes are extracted from mangrove bark. Leaves have been used in tea, medicine, livestock feed, and as a substitute for tobacco for smoking. In Florida, beekeepers have set up their hives close to mangroves in order to use the nectar in honey production.

Marine Life found in Mangroves

Many types of marine and terrestrial life utilize mangroves. Animals inhabit the mangrove's leafy canopy and waters underneath the mangrove's root system, and live in nearby tidal waters and mudflats.

In the U.S., larger species found in mangroves include reptiles such as the American crocodile and American alligator; sea turtles including the hawksbill, Ridley, green and loggerhead; fish such as snapper, tarpon, jack, sheepshead, and red drum; crustaceans such as shrimp and crabs; and coastal and migratory birds such as pelicans, spoonbills and bald eagles. In addition, less-visible species such as insects and crustaceans live among the roots and branches of the mangrove plants.

Threats to Mangroves

- Natural threats to mangroves include hurricanes, root clogging from increased water turbidity, and damage from boring organisms and parasites.

- Human impacts on mangroves have been severe in some places, and include dredging, filling, diking, oil spills, and runoff of human waste and herbicides. Some coastal development results in total loss of habitat.

Conservation of mangroves is important for the survival of mangrove species, humans and also for the survival of two other habitats – coral reefs and seagrass beds.

Estuary

An estuary is a partially enclosed body of water formed where freshwater from the land meets and mixes with saltwater from the ocean.

Estuaries vary in size and can also be termed bays, lagoons, harbours, inlets, sounds, wetlands and swamps.

- Estuaries are unique environments to which plants and animals have specially adapted.

- Transition from land to sea and fresh water to salt water

- Estuaries are protected from ocean forces by reefs, barrier islands, headlands and deltas.

- Estuaries transport and trap nutrients and sediment through the combined action of freshwater flow, wind, waves and tidal action.

- Some examples of estuaries in New Zealand include the Manukau Harbour, Raglan, Tairua, Avon-Heathcote and the fjords on the west coast of the South Island.

Formation of Estuaries

- Sea level has slowly risen over the last 15 000 years remaining stable over the last 6000 years. As the sea rose it drowned river valleys and filled glacial troughs. Once formed, estuaries make good sediment traps, filling with sediment from both the land and the sea. Sediment from the land include muds and clays delivered by rivers while sediments from the sea are usually clean sands pushed into the estuary by waves and tidal currents. Sediment can also come from shoreline erosion, windblown sediment and shell production.

- Estuaries have many shapes and sizes depending on how they were formed. In New Zealand about three hundred and one estuarine systems have been identified ranging from a few hectares to 15000 hectares. 164 are bar built estuaries, 56 are drowned river valleys, 65 are lagoons and 16 are fjords. We have an example of every type of estuary. This variety is due to varied geology, rainfall, land use, coastal wave climate and different basin shapes and the degree of infilling.

Types of Estuaries

Bar-built Estuaries

Bar-built estuaries form when a shallow lagoon or bay is protected from the ocean by a sand bar, sand delta or barrier island. Examples of these are found along the west coast of the North Island, eastern Coromandel and the Avon Heathcoate Estuary in Christchurch.

Tectonic Estuaries

Tectonic estuaries are caused by the folding or faulting of land surfaces. These estuaries are found along major fault lines, like the Manukau Harbour in Auckland.

Coastal Plain Estuaries

Coastal plain estuaries are formed by the sea level rising and filling an existing river valley. Examples of this are the Okura estuary, North Auckland.

Fjords and Rias

Fjords and rias are U-shaped valleys formed by glacial action. Fjords are found in areas with long histories of glacier activity, seen along the west coast of the South Island. A fjord has a sill and a ria does not.

Examples of Estuaries

- Barrier enclosed lagoons e.g. Tairua
- River mouth estuaries e.g. Mokau
- Coastal embayments e.g. Coromandel Harbour
- Drowned river valleys e.g. Raglan
- A semi enclosed bay e.g. Firth of Thames

Estuarine Ecosystems

- These are areas where both ocean and land contribute to a unique ecosystem.

- A basic feature is the instability of an estuary due to the ebb and flood of the tide.

- Plant and animal wastes are washed away, sediment is shifted and fresh and salt water are mixed.

- Estuaries provide a calm refuge from the open sea for millions of plants and animals.

- The diversity of habitats enclosed in estuaries supports enormous abundance and diversity of species e.g. fish, shellfish, lobsters, marine worms, reeds, seagrasses, mangroves, algae, and phytoplankton.

- Visiting species include birds which roost and feed, pelagic fish to spawn and use as nurseries.

- Estuaries are among the most productive environments on earth.

- 4 times more productive in plant matter than a rye grass pasture and 20 times more productive than the open sea. Extremely rich in organic matter and nutrients.

- Photosynthesis occurs throughout the water column and on the sediment surface – very productive.

- The margins of the estuary contain the food webs important producers e.g. algae, eelgrass, rushes and mangroves providing a huge amount of organic matter. Marshes and mangroves produce up to ten tones of plant detritus per hectare per year – considered organic factories.

- Sediments are important as they store organic matter and are the site of microbial activity.

- Micro organisms decompose complex organic compounds into useable forms – ammonia, nitrates and phosphates. This becomes the basic food for primary consumers like crabs, shellfish, snails, and marine worms. These burrowing invertebrates – especially sediment feeders continually turn over the deposited material.

- Filter feeders such as cockles and pipis continually add faecal deposits to the sediment.

- Visiting animals from land, sea and fresh water use the estuary for feeding, breeding, spawning and as nurseries for their young. Food is abundant and easy to access because of the shallow water attracting many types of birds including gulls, ducks and wading birds.

- 20 species of bird visit the Avon-Heathcote estuary in Christchurch.

- At least 30 species of commercial fish use New Zealand estuaries at some stage in their life cycle. Include snapper, flounder, mullet, rock fish, sole, kahawai, trevally, parore, red cod and gurnard. Freshwater eels, salmon,and whitebait migrate through estuaries at least twice in their life cycle.

- Many species of shark use estuaries as pupping grounds. They bear their young in the estuary. These young use the estuary for food and shelter until they are ready to head out to the ocean.

- Plants and animals have adapted specially for the different habitats of this unique ecosystem.

Estuarine Habitats

Estuaries enclose a diverse range of habitats from subtidal areas to intertidal areas. These include:

- Sheltered upper estuary mangroves, seagrass beds and marshes
- Highly energetic beaches on the ocean side of the estuary
- Rocky reefs
- Wave built bars in estuary mouths
- Deep estuarine channels where swift tidal currents flow
- Shallow open salt water and fresh water
- River deltas
- Tidal pools
- Muddy fringing marshes
- Mid-estuary sand banks
- Intertidal flats
- Estuarine beaches

Mangrove Habitat

Mangroves provide coastal protection, stabilize mudflats and provide important habitats for a range of animals. They are on of the worlds most productive vegetation types. The leaves they shed form a nutrient compost providing a valuable food source for many organisms. Mangrove communities include shellfish, worms, crabs, shrimp and other crustaceans living on and in the sediment around the trees. Barnacles, snails and oysters live on the trunks and aerial roots and spiders, insects and birds live amongst the leaves and branches. Mangroves are one of the starting points in estuarine food webs. Mangroves have become protected habitats in many countries including New Zealand showing a change in people's perception.

Estuarine Nutrient Cycling

Nutrients are taken up by plants or recycled by sediments. Nutrients are controlled by inputs from land, plant biomass and tidal flow. Estuaries are important sites of nutrient recycling in the coastal environment. Estuaries have a continuous source of nutrients from their freshwater inputs. These freshwater inputs often have additional nutrients compared to the natural situation because they receive wastes from cities towns and farms. The balance of nutrients in an estuary depends on the amount of input from the land, how much is taken up by plants or re-cycled in the sediments within the estuary and how much is exported or imported to and from the ocean in the tidal water.

Nutrients within the estuary are used by plants such as algae in the water, seagrasses, seaweeds and mangroves and they enter other parts of the food web when plants are grazed on by invertebrates. When the plants, or parts of them die, nutrients are re-cycled within the estuary by the actions of invertebrates and bacteria.

Some nutrients are essential to support the productivity of estuaries. Too many nutrients however, can alter the balance of nutrient cycling and lead to undesirable plant growths such as blooms of phytoplankton or seaweeds.

Estuarine Food Webs

Plants are grazed and filtered by invertebrates like snails, cockles and oysters. These are eaten by juvenile and other small fish which may be hunted by larger fish like snapper, kingfish. rays and even sharks.

Importance of Estuaries

- Ecological value.
- Economic value.
- Cultural value.

Ecological Value

- Estuaries are one of the most productive ecosystems on earth.
- They maintain water quality through natural filtration as microbes break down organic matter and sediments bind pollutants. Wetlands that fringe many estuaries also have other valuable functions. Water draining from the land carries sediments, nutrients, and other pollutants. Much of the sediments and pollutants are filtered out as the water flows through these fringing marshes. This creates cleaner and clearer water, which benefits both people and marine life. Wetland soils and plants like mangroves, sea grasses and reeds also act as a natural buffer between the land and ocean, absorbing floodwaters from land and storm surges from the ocean.
- They help maintain biodiversity by providing a diverse range of unique habitats that are critical for the survival of many species. Thousands of birds, mammals, fish and other wild-life use estuaries as places to live feed and reproduce. Migratory birds use estuaries to rest and gain food during their journeys. Many species of fish and shellfish use estuaries as nurseries to spawn and allow juveniles to grow.

Economic Value

- They are tourist attractions.
- Used for transport and industry.
- They have ecological importance to commercial fisheries.
- Maintain water quality which benefits both people and marine life
- Natural buffer between the land and ocean, absorbing floodwaters and storm surges.
- They help maintain biodiversity by providing a diverse range of unique habitats that are critical for the survival of many species.

- Many species of commercially important fish and shellfish use estuaries as nurseries to spawn and allow juveniles to grow.

Cultural Value

- People value estuaries for recreation, scientific knowledge, education, aesthetic and traditional practices. Boating, fishing, swimming, surfing, and bird watching are just a few of the numerous recreational activities people enjoy in estuaries. Unique estuarine habitats makes them valuable laboratories for scientists and students. Estuaries also provide a great deal of aesthetic enjoyment for the people who live, work, or use them for recreation.

- Estuaries were a resource sought after by Maori. Timber for building materials, rongoa (medicine), harakeke (flax) for weaving, kai (food) which included birds, fish, rats, eels, shellfish, humans. To make full advantage of this plentiful resource, Kainga, which were unfortified villages were established near estuaries.

- Eels were an important part of the Maori diet. They were caught by hand with a bob or eel pot. The bob was a huhu grub or worm attaches to a string or flax. The eels teeth were tangled in the fibres and the eel would then be landed. Camps were set up in favourable spots during the autumn spawning migration. The large number of eels caught were filleted and dried in the sun.

- Estuaries provide us with numerous resources upon which a dollar value cannot easily be placed. They offer benefits and services which greatly improve our lifestyle. Estuaries are an irreplaceable natural resource that must be managed carefully for the mutual benefit of all who enjoy and depend on them.

Conservation of Estuaries

- Estuaries are the preferred site for human settlement. New Zealand has many examples including Invercargill, Dunedin, Christchurch, Nelson, Wellington, Napier, Tauranga, Auckland, and Whangarei.

- Estuaries receive the final impact of human activities throughout the catchment. This means the use of the surrounding land and the rivers draining into them. Rubbish dumping, pollution, reclamation, trampling by stock, urban and industrial development and recreational pressures are some of the immediate impacts.

- Indirect threats – human induced erosion may cause accelerated siltation affecting plant and animal communities.

- Infilling can be accelerated by poorly managed urban development or forest clearance which increases the runoff. This can increase the turbidity of the water, accelerate sedimentation and spread of silts, muds and clay throughout the estuary. Sediments trap pollutants like fertilizer run off, petroleum by-products, persistent pesticides and heavy metals. Where fertilizer runoff is trapped in the sediment the process of organic decomposition by microbes speeds up depleting oxygen to dangerously low levels. Toxic compounds like ammonia and hydrogen sulphide accumulate and can be harmful to other estuarine organisms. All this may decrease the health and diversity of habitats.

Fringing communities are the most productive areas of an estuary – they are also the most sheltered and shallow parts, which makes them ideal targets for land reclamation and dredging. This may alter the entire balance of the food web.

Kelp Forest

Kelp Forests are underwater ecosystems formed in shallow water by the dense growth of several different species known as kelps. Though they look very much like plants, kelps are actually extremely large brown algae. Some species can reach heights (underwater) of 150 feet (45 m), and under ideal physical conditions, kelp can grow 18 inches (45 cm) in a single day. As a result of this incredible growth, kelp forests can develop very quickly in areas that they did not previously exist.

Kelp thrives in cold, nutrient-rich waters. Because kelp attaches to the seafloor and eventually grows to the water's surface and relies on sunlight to generate food and energy, kelp forests are always coastal and require shallow, relatively clear water. Generally speaking, kelps live further from the tropics than coral reefs, mangrove forests, and warm-water seagrass beds, so kelp forests do not overlap with those systems. Like those systems, though, kelp forests provide important three-dimensional, underwater habitat that is home to hundreds or thousands of species of invertebrates, fishes, and other algae. Some species aggregate and spawn in kelp forests or utilize these areas as juvenile nursery habitat. Large predatory species of sharks and marine mammals are known to hunt in the long corridors that form in kelp forests between rows of individual plants.

Though kelp forests are important ecosystems wherever they occur, they are more dynamic than the other systems mentioned above. In other words, they can disappear and reappear based on the oceanographic conditions and the population sizes of their primary herbivores. Warmer than normal summers and seasonal changes to currents that bring fewer nutrients to kelp forests (both sometimes occurring naturally) combine to weaken kelps and threaten their survival in some years. Strong individual storms can wipe out large areas of kelp forest, by ripping the kelp plants from the seafloor. Large gatherings of sea urchins (a primary herbivore in kelp forests) can prevent kelp plants from growing large enough to form forests. The cycle between these so called "urchin barrens" and well-developed kelp forests is a well-studied phenomenon in regions that are favorable for forest formation. Each of these natural alterations to kelp forest density or total area affects the community of invertebrates and fishes that live in this ecosystem. Population sizes of many of these species (including some that are commercially important food species) depend on the success of kelp growth each year.

Kelp

Kelp might look like a tree but, really, it is a kind of large, brown algae and algae are members of the Protista kingdom. Growing up from the ocean floor about 2–30 meters, and as much as 20–30 cm above the ocean's surface.

Kelp does not have roots. Instead, it is secured by holdfasts that lock onto substrates made of rock, or cobble. Although it functions in a way like a root, holdfasts do not absorb nutrients. Its backbone, or stipe, grows upward from the holdfast; attached to the stipe are leaves or fronds, where photosynthesis takes place.

Gas bladders (pneumatocysts) keep the top parts of the kelp afloat. Two kinds of kelp grow on the eastern Pacific coast: giant kelp (Macrocystis pyrifera) and bull kelp (Nereocystis leutkeana). Giant kelp, a perennial, has a pneumatocyst on each blade. Giant kelp has been known to live as much as seven years. Bull kelp, on the other hand, is an annual plant and has only one pneumatocyst holding up its many blades.

Both kelp types live their lives in two stages: first, as spores, which are released from its parent as male or female plants, which produce sperm or eggs, fertilize and, then, grow into the kelp's second stage — as a mature plant.

Rich in biodiversity, kelp forests grow along rocky shorelines, mostly on the Pacific coast, from Alaska to Baja, California. Alaska is home to three types of kelp: Macrocystis (two kinds, one of which is giant kelp), Nereocystis luetkeana (bull kelp) and Alaria fistulosa.

Kelp forests grow best in nutrient-rich, clear waters whose temperatures are between 42–72 degrees F (5–20 degrees C). The water must be clear so that sunlight can reach the ocean floor where the kelp life begins. If the water is too warm (warmer than 20 degrees), the kelp does not thrive as well. Its most successful growth is in geographic areas of upwelling and where the waters are always high in nutrients and always cold.

For kelp to survive it must be anchored to strong substrate, otherwise it will be yanked loose during storms. But, it also needs a moderate current. Kelp grows quickly, sometimes as much as 30–60 centimeters/day. Their life cycle varies, depending on the species. Some live for one year only; others have a longer lifespan.

Kelp forests are home to many different species, including fish, sea urchins and other marine animals, invertebrates, such as snails, and sea otters. For part of their lifetime, kelp forests in Alaska are home to 20 or more species of fish, which are attracted to the kelp by the food supply. Some fish, such as herring and Atka mackerel, spawn in kelp beds.

Mammals, such as sea lions and whales, are also likely to take up residence. Each part of the kelp - at the ocean floor, in the middle and above the water in the canopy formed by the fronds - provides a home to other species.

Sea otters and sea urchins are particularly important to the health and stability of kelp forests. Sea urchins like kelp - a lot - and, if their populations get too big, they may graze a kelp forest to death; if they remain, they may also inhibit the kelp's ability to grow again in that area. Sea otters like sea urchins - a lot - which is a good thing because they can keep the urchin population under control, thus helping to preserve the kelp forests.

Ecosystem Architecture

Rockfish swimming around giant kelp

A kelp forest off of the coast
of Anacapa Island, California

A diver in a kelp forest off
the coast of California

Giant kelp uses gas-filled floats to keep the plant
suspended, allowing the kelp blades near the ocean
surface to capture light for photosynthesis.

The architecture of a kelp forest ecosystem is based on its physical structure, which influences the associated species that define its community structure. Structurally, the ecosystem includes three guilds of kelp and two guilds occupied by other algae:

- Canopy kelps include the largest species and often constitute floating canopies that extend to the ocean surface (e.g., *Macrocystis* and *Alaria*).

- Stipitate kelps generally extend a few meters above the sea floor and can grow in dense aggregations (e.g., *Eisenia* and *Ecklonia*).

- Prostrate kelps lie near and along the sea floor (e.g., *Laminaria*).

- The benthic assemblage is composed of other algal species (e.g., filamentous and foliose functional groups, articulated corallines) and sessile organisms along the ocean bottom.

- Encrusting coralline algae directly and often extensively cover geologic substrate.

Multiple kelp species often co-exist within a forest; the term understory canopy refers to the stipitate and prostrate kelps. For example, a *Macrocystis* canopy may extend many meters above the seafloor towards the ocean surface, while an understory of the kelps *Eisenia* and *Pterygophora* reaches upward only a few meters. Beneath these kelps, a benthic assemblage of foliose red algae may occur. The dense vertical infrastructure with overlying canopy forms a system of microenvironments similar to those observed in a terrestrial forest, with a sunny canopy region, a partially shaded middle, and darkened seafloor. Each guild has associated organisms, which vary in their levels of dependence on the habitat, and the assemblage of these organisms can vary with kelp morphologies. For example, in California, *Macrocystis pyrifera* forests, the nudibranch *Melibe leonina*, and skeleton shrimp *Caprella californica* are closely associated with surface canopies; the kelp perch *Brachyistius frenatus*, rockfish *Sebastes* spp., and many other fishes are found within the stipitate understory; brittle stars and turban snails *Tegula* spp. are closely associated with the kelp holdfast, while various herbivores, such as sea urchins and abalone, live under the prostrate canopy; many seastars, hydroids, and benthic fishes live among the benthic assemblages; solitary corals, various gastropods, and echinoderms live over the encrusting coralline algae. In addition, pelagic fishes and marine mammals are loosely associated with kelp forests, usually interacting near the edges as they visit to feed on resident organisms.

Trophic Ecology

Sea urchins like this purple sea urchin can damage kelp forests by chewing through kelp holdfasts

The jeweled top snail *Calliostoma annulatum* grazing on a blade of giant kelp

The sea otter is an important predator of sea urchins.

Classic studies in kelp forest ecology have largely focused on trophic interactions (the relationships between organisms and their food webs), particularly the understanding and top-down trophic processes. Bottom-up processes are generally driven by the abiotic conditions required for primary producers to grow, such as availability of light and nutrients, and the subsequent transfer of energy to consumers at higher trophic levels. For example, the occurrence of kelp is frequently correlated with oceanographic upwelling zones, which provide unusually high concentrations of nutrients to the local environment. This allows kelp to grow and subsequently support herbivores, which in turn support consumers at higher trophic levels. By contrast, in top-down processes, predators limit the biomass of species at lower trophic levels through consumption. In the absence of predation, these lower-level species flourish because resources that support their energetic requirements are not limiting. In a well-studied example from Alaskan kelp forests, sea otters (Enhydra lutris) control populations of herbivorous sea urchins through predation. When sea otters are removed from the ecosystem (for example, by human exploitation), urchin populations are released from predatory control and grow dramatically. This leads to increased herbivore pressure on local kelp stands. Deterioration of the kelp itself results in the loss of physical ecosystem structure and subsequently, the loss of other species associated with this habitat. In Alaskan kelp forest ecosystems, sea otters are the keystone species that mediates this trophic cascade. In Southern California, kelp forests persist without sea otters and the control of herbivorous urchins is instead mediated by a suite of predators including lobsters and large fishes, such as the California sheephead. The effect of removing one predatory species in this system differs from Alaska because redundancy exists in the trophic levels and other predatory species can continue to regulate urchins. However, the removal of multiple predators can effectively release urchins from predator pressure and allow the system to follow trajectories towards kelp forest degradation. Similar examples exist in Nova Scotia, South Africa, Australia and Chile. The relative importance of top-down versus bottom-up control in kelp forest ecosystems and the strengths of trophic interactions continue to be the subject of considerable scientific investigation.

The transition from macroalgal (i.e. kelp forest) to denuded landscapes dominated by sea urchins (or 'urchin barrens') is a widespread phenomenon, often resulting from trophic cascades like those described above; the two phases are regarded as alternative stable states of the ecosystem. The recovery of kelp forests from barren states has been documented following dramatic perturbations, such as urchin disease or large shifts in thermal conditions. Recovery from intermediate states of deterioration is less predictable and depends on a combination of abiotic factors and biotic interactions in each case.

Though urchins are usually the dominant herbivores, others with significant interaction strengths include seastars, isopods, kelp crabs, and herbivorous fishes. In many cases, these organisms feed on kelp that has been dislodged from substrate and drifts near the ocean floor rather than expend energy searching for intact thalli on which to feed. When sufficient drift kelp is available, herbivorous grazers do not exert pressure on attached plants; when drift subsidies are unavailable, grazers directly impact the physical structure of the ecosystem. Many studies in Southern California have demonstrated that the availability of drift kelp specifically influences the foraging behavior of sea urchins. Drift kelp and kelp-derived particulate matter have also been important in subsidizing adjacent habitats, such as sandy beaches and the rocky intertidal.

Patch Dynamics

Another major area of kelp forest research has been directed at understanding the spatial-temporal patterns of kelp patches. Not only do such dynamics affect the physical landscape, but they also affect species that associate with kelp for refuge or foraging activities. Large-scale environmental disturbances have offered important insights concerning mechanisms and ecosystem resilience. Examples of environmental disturbances include:

- Acute and chronic pollution events have been shown to impact southern California kelp forests, though the intensity of the impact seems to depend on both the nature of the contaminants and duration of exposure. Pollution can include sediment deposition and eutrophication from sewage, industrial byproducts and contaminants like PCBs and heavy metals (for example, copper, zinc), runoff of organophosphates from agricultural areas, anti-fouling chemicals used in harbors and marinas (for example, TBT and creosote) and land-based pathogens like fecal coliform bacteria.

- Catastrophic storms can remove surface kelp canopies through wave activity, but usually leave understory kelps intact; they can also remove urchins when little spatial refuge is available. Interspersed canopy clearings create a seascape mosaic where sunlight penetrates deeper into the kelp forest and species that are normally light-limited in the understory can flourish. Similarly, substrate cleared of kelp holdfasts can provide space for other sessile species to establish themselves and occupy the seafloor, sometimes directly competing with juvenile kelp and even inhibiting their settlement.

- El Niño-Southern Oscillation (ENSO) events involve the depression of oceanographic thermoclines, severe reductions of nutrient input, and changes in storm patterns. Stress due to warm water and nutrient depletion can increase the susceptibility of kelp to storm damage and herbivorous grazing, sometimes even prompting phase shifts to urchin-dominated landscapes. In general, oceanographic conditions (that is, water temperature, currents) influence the recruitment success of kelp and its competitors, which clearly affect subsequent species interactions and kelp forest dynamics.

- Overfishing higher trophic levels that naturally regulate herbivore populations is also recognized as an important stressor in kelp forests. The drivers and outcomes of trophic cascades are important for understanding spatial-temporal patterns of kelp forests.

In addition to ecological monitoring of kelp forests before, during, and after such disturbances, scientists try to tease apart the intricacies of kelp forest dynamics using experimental manipulations. By working on smaller spatial-temporal scales, they can control for the presence or absence of specific biotic and abiotic factors to discover the operative mechanisms. For example, in southern Australia, manipulations of kelp canopy types demonstrated that the relative amount of *Ecklonia radiata* in a canopy could be used to predict understory species assemblages; consequently, the proportion of *E. radiata* can be used as an indicator of other species occurring in the environment.

Human Use

Kelp forests have been important to human existence for thousands of years. Indeed, many now theorize that the first colonization of the Americas was due to fishing communities following the Pacific kelp forests during the last ice age. One theory contends that the kelp forests that would have stretched from northeast Asia to the American Pacific coast would have provided many benefits to ancient boaters. The kelp forests would have provided many sustenance opportunities, as well as acting as a type of buffer from rough water. Besides these benefits, researchers believe that the kelp forests might have helped early boaters navigate, acting as a type of "kelp highway". Theorists also suggest that the kelp forests would have helped these ancient colonists by providing a stable way of life and preventing them from having to adapt to new ecosystems and develop new survival methods even as they traveled thousands of miles. Modern economies are based on fisheries of kelp-associated species such as lobster and rockfish. Humans can also harvest kelp directly to feed aquaculture species such as abalone and to extract the compound alginic acid, which is used in products like toothpaste and antacids. Kelp forests are valued for recreational activities such as SCUBA diving and kayaking; the industries that support these sports represent one benefit related to the ecosystem and the enjoyment derived from these activities represents another. All of these are examples of ecosystem services provided specifically by kelp forests.

Threats and Management

The nudibranch Melibe leonina on a Macrocystis frond (California):
Marine protected areas are one way to guard kelp forests as an ecosystem.

Given the complexity of kelp forests – their variable structure, geography, and interactions – they pose a considerable challenge to environmental managers. Extrapolating even well studied trends to the future is difficult because interactions within the ecosystem will change under variable conditions, not all relationships in the ecosystem are understood, and the nonlinear thresholds to transitions are not yet recognized. With respect to kelp forests, major issues of concern include marine pollution and water quality, kelp harvesting and fisheries, invasive species, and climate change. The most pressing threat to kelp forest preservation may be the overfishing of coastal ecosystems, which by removing higher trophic levels facilitates their shift to depauperate urchin barrens. The maintenance of biodiversity is recognized as a way of generally stabilizing ecosystems and their services through mechanisms such as functional compensation and reduced susceptibility to foreign species invasions.

In many places, managers have opted to regulate the harvest of kelp and/or the taking of kelp forest species by fisheries. While these may be effective in one sense, they do not necessarily protect the entirety of the ecosystem. Marine protected areas (MPAs) offer a unique solution that encompasses not only target species for harvesting, but also the interactions surrounding them and the local environment as a whole. Direct benefits of MPAs to fisheries (for example, spillover effects) have been well documented around the world. Indirect benefits have also been shown for several cases among species such as abalone and fishes in Central California. Most importantly, MPAs can be effective at protecting existing kelp forest ecosystems and may also allow for the regeneration of those that have been affected.

Seagrass

Seagrasses are found in shallow salty and brackish waters in many parts of the world, from the tropics to the Arctic Circle. Seagrasses are so-named because most species have long green, grass-like leaves. They are often confused with seaweeds, but are actually more closely related to the flowering plants that you see on land. Seagrasses have roots, stems and leaves, and produce flowers and seeds. They evolved around 100 million years ago, and today there are approximately 72 different seagrass species that belong to four major groups. Seagrasses can form dense underwater meadows, some of which are large enough to be seen from space. Although they often receive little attention, they are one of the most productive ecosystems in the world. Seagrasses provide shelter and food to an incredibly diverse community of animals, from tiny invertebrates to large fish, crabs, turtles, marine mammals and birds. Seagrasses provide many important services to people as well, but many seagrasses meadows have been lost because of human activities. Work is ongoing around the world to restore these important ecosystems.

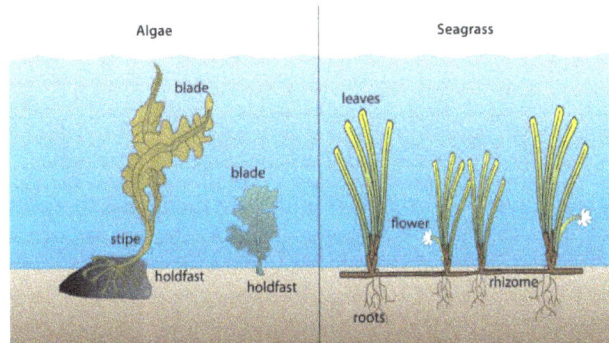

Algae or "seaweeds" (left) differ from seagrasses (right) in several ways. Algae on the seafloor have a holdfast and transport nutrients through the body by diffusion, while seagrasses are flowering vascular plants with roots and an internal transport system.

Even though seagrasses and seaweeds look superficially similar, they are very different organisms. Seagrasses belong to a group of plants called monocotyledons that include grasses, lilies and palms. Like their relatives, seagrasses have leaves, roots and veins, and produce flowers and seeds. Chloroplasts in their tissues use the sun's energy to convert carbon dioxide and water into sugar and oxygen for growth through the process of photosynthesis. Veins transport nutrients and water throughout the plant, and have little air pockets called lacunae that help keep the leaves buoyant and exchange oxygen and carbon dioxide throughout the plant. Like other flowering plants, their roots can absorb nutrients. Unlike flowering plants on land, however, they lack stomata—the tiny pores on leaves that open and close to control water and gas exchange. Instead, they have a thin cuticle layer, which allows gasses and nutrients to diffuse directly into and out of the leaves from the water. The roots and rhizomes (thicker horizontal stems) of seagrasses extend into the sediment of the seafloor and are used to store and absorb nutrients, as well as anchor the plants. In contrast, seaweeds (algae) are much simpler organisms. They have no flowers or veins, and their holdfasts simply attach to the bottom and are generally not specialized to take in nutrients. Scientists are studying what genes were lost and which were regained as seagrasses evolved from algae in the sea to plants on land, and then transitioned back to the sea. The entire genome of one seagrass, the eelgrass *Zostera marina*, was sequenced in 2016, helping us understand how these plants adapted to life in the sea, how they may respond to climate warming, and the evolution of salt tolerance in crop plants.

Seagrass Ecosystem

Seagrasses grow in salty and brackish (semi-salty) waters around the world, typically along gently sloping, protected coastlines. Because they depend on light for photosynthesis, they are most commonly found in shallow depths where light levels are high. Many seagrass species live in depths of 3 to 9 feet (1 to 3 meters), but the deepest growing seagrass (*Halophila decipiens*) has been found at depths of 190 feet (58 meters). While most coastal regions are dominated by one or a few seagrass species, regions in the tropical waters of the Indian and western Pacific oceans have the highest seagrass diversity with as many as 14 species growing together. Antarctica is the only continent without seagrasses.

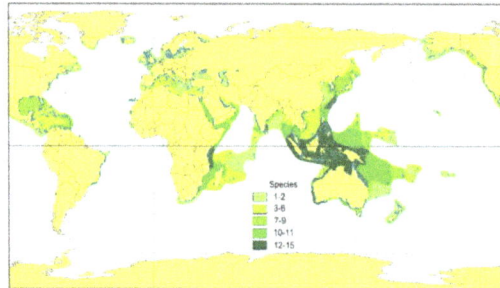

Seagrasses are found across the world, from the tropics to the Arctic.
Shades of green indicate the number of species reported for a given area.
The darker shades of green indicate more species are present.

Growth & Reproduction

Seagrasses grow both vertically and horizontally—their blades reach upwards and their roots down and sideways—to capture sunlight and nutrients from the water and sediment. They spread by two methods: asexual clonal growth and sexual reproduction.

Asexual Clonal Growth: Similar to grasses on land, seagrass shoots are connected underground by a network of large root-like structures called rhizomes. The rhizomes can spread under the sediment and send up new shoots. When this happens, many stems within the same meadow can actually be part of the same plant and will have the same genetic code—which is why it is called clonal growth. In fact, the oldest known plant is a clone of the Mediterranean seagrass *Posidonia oceanica*, which may be up to 200,000 years old, dating back to the ice ages of the late Pleistocene. In some seagrass species, a meadow can develop from a single plant in less than a year, while in slow-growing species like *Posidonia* it can take hundreds of years.

Sexual Reproduction: Seagrasses reproduce sexually like terrestrial grasses, but pollination for seagrasses is completed with the help of water. Male seagrass flowers release pollen from structures called stamens into the water. Seagrasses produce the longest pollen grains on the planet (up to 5mm long compared to under 0.1mm for land plants typically), and this pollen often collects into stringy clumps. The clumps are moved by currents until they land on the pistil of a female flower and fertilization takes place. There is also evidence that small invertebrates, such as amphipods (tiny shrimp-like crustaceans) and polychaetes (marine worms), feed on the pollen of one seagrass (*Thalassia testudinum*), which could help to fertilize the flowers in a way similar to how insects pollinate flowers on land.

Self-pollination happens in some grass species, which can reduce genetic variation. Individual seagrass plants avoid this by producing only male or female flowers, or by producing the male and female flowers at different times. Just like land grasses, fertilized seagrass flowers develop seeds. Seagrass seeds are neutrally buoyant and can float many miles before they settle onto the soft sea-floor and germinate to form a new plant. A few seagrass species such as the surfgrass *Phylospadix* can settle and live on rocky shores. Animals that eat seagrass seeds—including fish and turtles—may incidentally aid with their dispersal and germination if the seeds pass through their digestive tracks and remain viable.

Biodiversity

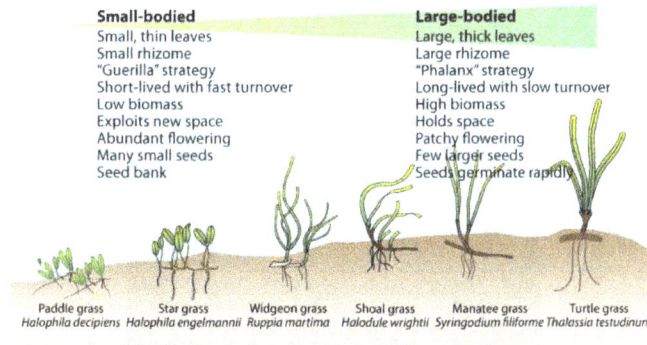

Small-bodied	Large-bodied
Small, thin leaves	Large, thick leaves
Small rhizome	Large rhizome
"Guerilla" strategy	"Phalanx" strategy
Short-lived with fast turnover	Long-lived with slow turnover
Low biomass	High biomass
Exploits new space	Holds space
Abundant flowering	Patchy flowering
Many small seeds	Few larger seeds
Seed bank	Seeds germinate rapidly

Paddle grass	Star grass	Widgeon grass	Shoal grass	Manatee grass	Turtle grass
Halophila decipiens	Halophila engelmannii	Ruppia martima	Halodule wrightii	Syringodium filiforme	Thalassia testudinum

Different shapes and sizes of seagrass species

The 72 species of seagrasses are commonly divided into four main groups: Zosteraceae, Hydrocharitaceae, Posidoniaceae and Cymodoceaceae. Their common names, like eelgrass, turtle grass, tape grass, shoal grass, and spoon grass, reflect their many shapes and sizes and roles in marine ecosystems. Seagrasses range from species with long flat blades that look like ribbons to fern or paddle-shaped leaves, cylindrical or spaghetti blades, or branching shoots. The tallest seagrass species—*Zostera caulescens*—was found growing to 35 feet (7 meters) in Japan. Some seagrass species are quick growing while others grow much more slowly. These distinct structures and growth forms affect how seagrasses influence their environment and what species live in the habitats they create.

Ecosystem Benefits

Seagrasses are often called foundation plant species or ecosystem engineers because they modify their environments to create unique habitats. These modifications not only make coastal habitats more suitable for the seagrasses themselves, but also have important effects on other animals and provide ecological functions and a variety of services for humans.

Seagrasses have been used by humans for over 10,000 years. They've been used to fertilize fields, insulate houses, weave furniture, thatch roofs, make bandages, and fill mattresses and even car seats. But it's what they do in their native habitat that has the biggest benefits for humans and the ocean. Seagrasses support commercial fisheries and biodiversity, clean the surrounding water and help take carbon dioxide out of the atmosphere. Because of these benefits, seagrasses are believed to be the third most valuable ecosystem in the world (only preceded by estuaries and wetlands). One hectare of seagrass (about two football fields) is estimated to be worth over $19,000 per year, making them one of the most valuable ecosystems on the planet.

Key Services

Modification of the Physical Environment

Seagrasses are known as the "lungs of the sea" because one square meter of seagrass can generate 10 liters of oxygen every day through photosynthesis. Seagrass leaves also absorb nutrients and slow the flow of water, capturing sand, dirt and silt particles. Their roots trap and stabilize the sediment, which not only helps improve water clarity and quality, but also reduces erosion and buffers coastlines against storms. Seagrasses can further improve water quality by absorbing nutrients in runoff from the land. In nutrient poor regions, the seagrass plants themselves help nutrient cycling by taking up nutrients from the soil and releasing them into the water through their leaves, acting as a nutrient pump.

An Australian Giant Cuttlefish (*Sepia apama*) crosses a seagrass bed.

Creation of Living Habitat

Seagrasses don't just provide shelter for free-swimming animals,
but also are a habitat for non-moving organisms, such as these sea anemones.

Seagrasses are often called nursery habitats because the leafy underwater canopy they create provides shelter for small invertebrates (like crabs and shrimp and other types of crustaceans), small fish and juveniles of larger fish species. Many species of algae and microalgae (such as diatoms), bacteria and invertebrates grow as "epiphytes" directly on living seagrass leaves, much like lichens and Spanish moss grow on trees. Other invertebrates grow nestled between the blades or in the sediments—such as sponges, clams, polychaete worms and sea anemones. The accumulation of smaller organisms amongst and on the seagrass blades, as well as the seagrass itself, attracts bigger animals. As a result, seagrasses can be home to many types of fish, sharks, turtles, marine mammals (dugongs and manatees), mollusks (octopus, squid, cuttlefish, snails, bivalves), sponges, crustaceans (shrimp, crabs, copepods, isopods and amphipods) polychaete worms, sea urchins and sea anemones—and the list goes on.

Some of these organisms are permanent residents in seagrass meadows, while others are temporary visitors. A single acre of seagrass can support upwards of 40,000 fish and 50 million small invertebrates, and there are often tens to hundreds more animals in a seagrass bed compared to adjacent bare sandy areas. A number of the species that depend on seagrasses are important for commercial and recreational fisheries. In fact, in all regions of the world fishermen will specifically seek out seagrass beds for their abundance of fish. It is because of the wide variety of different species that live amongst the grasses that seagrass beds often form important "biodiversity hotspots." Not only do seagrasses support a diversity of marine life, but populations of a given seagrass species can themselves be very genetically diverse and this diversity itself is linked to higher animal abundances. Understanding how seagrass genotypic diversity does this is an active area of research.

Foundation of Coastal Food Webs

Adult green sea turtles spend most of their time grazing in seagrass meadows.

Seagrass beds are important feeding grounds for thousands of species around the world, and they support this diverse food web in three different ways. Some organisms—primarily large grazers like manatees, dugongs, green sea turtles and geese—eat the living leaves directly, and seagrass forms a major component of their diets. For example, an adult dugong eats about 64 to 88 pounds (28 to 40 kg) of seagrass a day, while an adult green sea turtle can eat about 4.5 pounds (2 kg) per day. Many of these large grazers are endangered, in large part because of habitat destruction and hunting, but once they were very common. It's estimated that before Europeans settled the Americas in the 1400's, the number of green sea turtles supported by seagrass meadows was 15 to 20 times the number and biomass of large hooved animals in the Serengeti Desert alive today. These abundant large grazers probably kept seagrass meadows cropped short like a putting green.

Small invertebrates, such as these crustaceans (left) and gastropods (right), can help keep seagrasses clean by consuming epiphytic algae

The epiphytic organisms growing on the surface of the seagrass blades provide other sources of food. Some epiphytic bacteria can extract nitrogen from the environment and make it available to larger animals. Small invertebrate mesograzers, such as crustaceans and snails, feed on epiphytes, and in doing so can help keep the seagrass clean, acting as mutualistic partners (or housekeepers) that promote seagrass growth. They are in turn consumed by larger crustaceans, fish and birds and are important links in the coastal food web. But, this partnership isn't always positive. Occasionally when some mesograzer species are at very high densities they can create thick masses of mucus and sediment tubes that block light to the seagrass leaves, and they can even eat the seagrass directly.

Dead seagrass leaves also play an important role in coastal ecosystems. When the leaves die, they decay on the sediment or are washed onto the beach, supporting a diverse community of decomposers that thrive on rotting material. Some of these living and dead seagrass blades are also washed to other areas of the ocean, feeding organisms in ecosystems as far as the deep sea.

Blue Carbon

Atmospheric carbon is captured by coastal mangroves, seagrasses
and salt marshes at a rate five times faster than tropical forests.

Seagrasses are capable of capturing and storing a large amount of carbon from the atmosphere. Similar to how trees take carbon from the air to build their trunks, seagrasses take carbon from the water to build their leaves and roots. As parts of the seagrass plants and associated organisms die and decay, they can collect on the seafloor and become buried, trapped in the sediment. It has been estimated that in this way the world's seagrass meadows can capture up to 83 million metric tons of carbon each year. The carbon stored in sediments from coastal ecosystems including seagrass meadows, mangrove forests and salt marshes is known as "blue carbon" because it is stored in the sea. While seagrasses occupy only 0.1 percent of the total ocean floor, they are estimated to be responsible for up to 11 percent of the organic carbon buried in the ocean. One acre of seagrass can sequester 740 pounds of carbon per year (83 g carbon per square meter per year), the same amount emitted by a car traveling around 3,860 miles (6,212 km).

Threats & Conservation

Unfortunately, seagrasses are in trouble. Seagrass coverage is being lost globally at a rate of 1.5 percent per year. That amounts to about 2 football fields of seagrass lost each hour. It's estimated that 29 percent of seagrass meadows have died off in the past century. In a 2011 assessment, nearly one quarter of all seagrass species for which information was adequate to judge were threatened (endangered or vulnerable) or near threatened using the International Union for the Conservation of Nature (IUCN) Red List criteria. This is especially worrying because seagrass losses are projected to have severe impacts on marine biodiversity, the health of other marine ecosystems, and on human livelihoods. Additionally, some threatened marine species such as sea turtles and marine mammals live in seagrass habitats and rely on them for food. For every seagrass species there is on average more than one associated threatened marine species. In fact, the only marine plant listed as endangered in the United States is a seagrass (*Halophila johnsonii*) found in Florida.

Threats To Seagrasses

A strain of *Caulerpa taxifolia* escaped aquariums and has spread widely in the Mediterranean, replacing native plants and depriving marine life of food and habitat

Seagrasses are vulnerable to physical disturbances, such as wind-driven waves and storms. Some animals, such as skates and rays, disturb the rhizomes and roots of seagrasses, ripping up the seagrass as they forage for buried clams and other invertebrates. However, the direct and indirect effects of human activities account for most losses of seagrass beds in recent decades. Some fast growing seagrass meadows are able to rebound from disturbances, but many grow slowly over the course of centuries and are likely to be slow to recover and are thus most vulnerable.

Nutrients, such as those from fertilizers and pollution, wash off the land and into the water, causing algal blooms that block sunlight necessary for seagrass growth. Sediment washing into the water from agriculture and land development can also damage seagrass beds by both smothering the seagrass and blocking sunlight. Similarly, dredging can both directly remove seagrass plants and cause lower light levels because of increased amounts of sediments in the water. Boat anchors and propellers can leave "scars" in a seagrass bed—killing sections of the seagrass and fragmenting the habitat. This fragmentation of seagrass beds can increase erosion around the edges, as well as influence animal use and movement within the seagrass bed.

Disease has also devastated seagrasses. In the early 1930s, a large die-off of up to 90 percent of all eelgrass (*Zostera marina*) growing in temperate North America was attributed to a "wasting disease". This die-off was so severe that a small snail specialized to live on eelgrass went extinct as a result. The disease was caused by the slime mold-like protist, *Labyrinthula zosterae*, which also ravaged eelgrass populations in Europe. This disease still affects eelgrass populations in the Atlantic and has contributed to some recent losses, though none as catastrophic as in the 1930s. Eelgrass leaves that are weak or stressed are more susceptible to the disease, developing brown spots and lesions that reduce the plant's ability to photosynthesize, eventually killing the plant. Healthy plants are thought to be resistant to the disease, indicating importance of reducing other stressors like pollution. Lower seawater salinity may also increase susceptibility to the *Labyrinthula* pathogen.

Episodes of warm seawater temperatures can also damage seagrasses. Temperature affects how enzymes and metabolism work, influencing how organisms grow. Rising water temperatures tend to increase rates of seagrass respiration (using up oxygen) faster than rates of photosynthesis (producing oxygen), which makes them more susceptible to grazing by herbivores. Increased temperature also increases seagrass light requirements, influences how quickly seagrasses can take up nutrients in their environment, and can make seagrasses more susceptible to disease. Large eelgrass declines have been observed in the Chesapeake Bay in years in which water temperatures have persisted for several days above 30°C (86°F), the thermal limit for this species.

Removal of fish can also lead to seagrass death by disrupting important components of the food web. When large predators are removed, intermediate predators can become more abundant, and they in turn cause the decline of the smaller organisms that keep the blades of the seagrasses clean. This has been observed most strikingly in the Baltic sea with the disappearance of cod due to overfishing and corresponding increases in smaller fishes and crustaceans which limited epiphyte-grazing invertebrates, resulting in seagrass decline.

In addition to the small epiphytic algae, larger algae also compete with seagrasses, and introduced invasive seaweed species can displace native seagrass species. One important example is the invasion of *Caulerpa taxifolia*, a seaweed nicknamed "the killer algae." Released into the Mediterranean in the 1980s from aquaria, by 2000 it covered more than 131 square kilometers (50 square miles) of the Mediterranean coastline, overgrowing and replacing the native Neptune seagrass (*Posidonia oceanica*) and reducing the ecosystem's biodiversity. Since then, invasive Caulerpa has been found in California and southwestern Australia where eradication programs are in place to prevent its spread.

Protecting and Restoring Seagrass Beds

Neptune grass (*Posidonia oceanica*) is a slow-growing
and long-lived seagrass native to the Mediterranean.

Most management that protects seagrasses focuses on maintaining their biodiversity and the services these habitats provide for humans and ecosystems. There is no international legislation for seagrasses, and so protection typically occurs by local and regional agencies. Actions taken to help seagrasses include limiting damaging practices such as excessive trawling and dredging, runoff pollution and harmful fishing practices (such as dynamite or cyanide fishing).

There are also attempts to rebuild and restore seagrass beds, often by planting seeds or seedlings grown in aquaria, or transplanting adult seagrasses from other healthy meadows. Some of the most successful restoration stories come from the Chesapeake Bay and coastal Virginia in the Eastern United States where, through 2014, the Virginia Institute of Marine Science has seeded 456 acres with 7.65 million seagrass seeds. As of 2015, the seagrass *Zostera marina* has increased from these seeded plots to cover 6,195 acres. Seagrass restoration in Tampa Bay, Florida, has also experienced important success including improvements in water quality and the associated fish community. For restoration to work, it is critical that the causes of the original decline in seagrasses have been eliminated.

However, seagrass populations globally are still in trouble. Some simple steps everyone can take to help seagrasses and other marine habitats include: don't litter, limit the amount of fertilizer and pesticides you use, don't dump anything hazardous down the drain, be careful when boating by going slow and avoiding shallow areas, and support local conservation efforts.

Pelagic Zone

The pelagic zone is the area of the ocean outside of coastal areas. This is also called the open ocean. The open ocean lies over and beyond the continental shelf. It's where you'll find some of the biggest marine life species.

The sea floor (demersal zone) is not included in the pelagic zone.

Different Zones within the Pelagic Zone

The pelagic zone is separated into several subzones depending on water depth:

- Epipelagic zone (ocean surface to 200 meters deep). This is the zone in which photosynthesis can occur because light is available.

- Mesopelagic zone (200-1,000m) - This is also known as the twilight zone because light becomes limited. There is less oxygen available to organisms in this zone.

- Bathypelagic zone (1,000-4,000m) - This is a dark zone where water pressure is high and the water is cold (around 35-39 degrees).

- Abyssopelagic zone (4,000-6,000m) - This is the zone past the continental slope - the deep water just over the ocean bottom. This is also known as the abyssal zone.

- Hadopelagic zone (deep ocean trenches, greater than 6,000m) - In some places, there are trenches that are deeper than the surrounding ocean floor. These areas are the hadopelagic zone. At a depth of over 36,000 feet, the Mariana Trench is the deepest known point in the ocean.

Within these different zones, there can be a dramatic difference in available light, water pressure and the types of species you'll find there.

Marine Life Found in the Pelagic Zone

Thousands of species of all shapes and sizes live in the pelagic zone. You'll find animals that travel long distances and some that drift with the currents. There is a wide array of species here as this zone includes the entire ocean that is not either in a coastal area or the ocean bottom. Thus, the pelagic zone thus comprises the largest volume of ocean water in any marine habitat.

Life in this zone ranges from tiny plankton to the largest whales.

Plankton

Organisms include phytoplankton, which provides oxygen for us here on Earth and food for many animals. Zooplanktons such as copepods are found there and also are an important part of the oceanic food web.

Invertebrates

Examples of invertebrates that live in the pelagic zone include jellyfish, squid, krill, and octopus.

Vertebrates

Many large ocean vertebrates live in or migrate through the pelagic zone. These include cetaceans, sea turtles and large fish such as ocean sunfish, bluefin tuna, swordfish, and sharks.

While they don't live *in* the water, seabirds such as petrels, shearwaters, and gannets can often be found above, on and diving under the water in search of prey.

Challenges of the Pelagic Zone

This can be a challenging environment where species are affected by wave and wind activity, pressure, water temperature and prey availability. Because the pelagic zone covers a large area, prey may be scattered over some distance, meaning animals have to travel far to find it and may not feed as often as an animal in a coral reef or tide pool habitat, where prey is denser.

Some pelagic zone animals (e.g., pelagic seabirds, whales, sea turtles) travel thousands of miles between breeding and feeding grounds. Along the way, they face changes in water temperatures, types of prey, and human activities such as shipping, fishing, and exploration.

Surface Water

Surface water is any natural water that has not penetrated under the surface of the ground underneath. It is unlike ground-water, which is underground or has seeped under the surface of the earth. Rivers, lakes, oceans and wetlands are commonly known bodies of surface water. Surface water is lost through evaporation and regained through precipitation (rain) or recruited from ground-water sources.

Surface water that is not saline (salt) can also be lost by seeping into the ground, where it becomes ground-water, used by plants, mankind for life support, industrial purposes or agricultural purposes, or can enter the sea where it becomes saline water. For reasons of environmental health and safety, there is a need for increased management of both surface and ground-water. This is because to ensure good management of surface water means good management of ground water, as they are interrelated in the water cycle. Surface water is a resource that is being depleted in our modern world due to over-pumping of rivers, lakes, etc.

Properties of Surface Water

There are many important properties of surface water, including temperature, saltiness (also called salinity), turbidity, and levels of dissolved nutrients, such as oxygen and carbon dioxide. These factors all affect climate and the biodiversity in and around a body of water.

- Temperature, scientifically, is the average kinetic energy of the molecules in a substance. The temperature of surface water is warmest at the top, and it gets cooler as you go deeper. The deep oceans, for example, are extremely cold, dark places. Surface water temperature varies much more by season as well. So, when we look at temperature, we tend to look at averages.

While we're talking about temperature, an interesting fact is that sea water doesn't freeze at 0 degrees Celsius: the salt in the water allows it to get colder before it does. Also, warmer water tends to have less oxygen, which can make it harder for some species of marine life to survive.

- Salinity is the saltiness of water. It measures the amount of dissolved sodium, potassium, and other salts in the water. Higher salinity leads to denser water, which has an impact on water currents around the world. Areas with a lot of evaporation have higher salinity and denser water, because when water evaporates, it leaves the salts behind. Here's a representation of sea surface salinity around the globe.

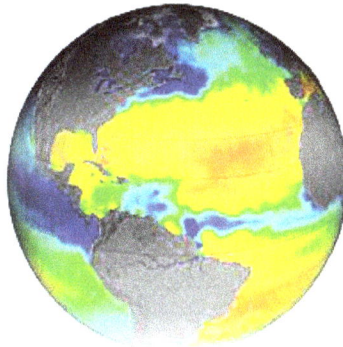

Deep Sea

Most people familiar with the oceans know about life only in the intertidal zone, where the water meets land, and the epipelagic zone, the upper sunlit zone of the open ocean. Though these zones contain an abundance of ocean life because sunlight is available for photosynthesis, they make up only a small fraction of the ocean biome. In fact, most of the ocean is cold, dark and deep. It is important to realize that photosynthesis occurs only down to about 100 - 200 m, and sunlight disappears altogether at 1,000 m or less, while the ocean descends to a maximum depth of about 11,000 m in the Mariana Trench.

To get an idea of how vast the ocean's depths are, consider that 79% of the entire volume of the earth's biosphere consists of waters with depths greater than 1,000 m. Until recently, the deep sea was largely unexplored. But advances in deep sea submersibles and image capturing and sampling technologies are increasing the opportunities for marine biologists to observe and uncover the mysteries of the deep ocean realm.

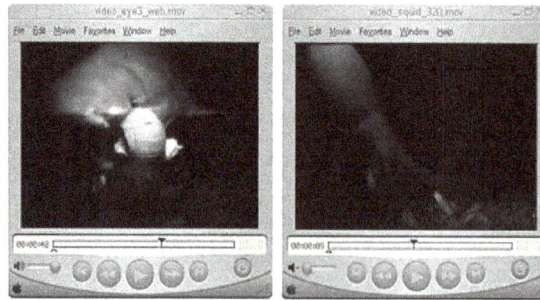

Deep sea research is vital because this area is such an enormous part of the biosphere. Despite its depth and distance, it is still our backyard in comparison to outer space. And yet, human exploration has revealed more detail about the surface of the moon and Mars that it has about the deep sea. Consider that hydrothermal vents and their unique organisms, which revolutionized our ideas about energy sources and the adaptability of life, were only discovered in 1977. There may be yet other life-altering discoveries to be found at the bottom of the ocean.

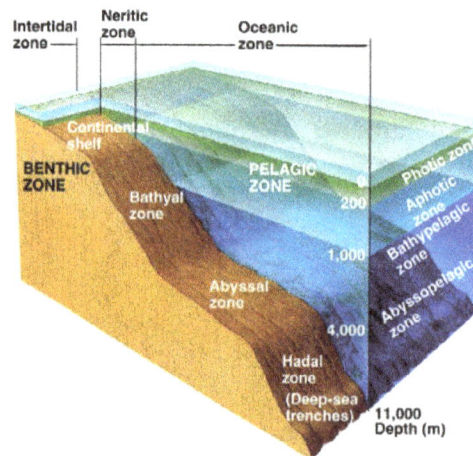

The oceans are divided into two broad realms; the pelagic and the benthic. Pelagic refers to the open water in which swimming and floating organisms live. Organisms living there are called the pelagos. From the shallowest to the deepest, biologists divide the pelagic into the epipelagic (less than 200 meters, where there can be photosynthesis), the mesopelagic (200 - 1,000 meters, the "twilight" zone with faint sunlight but no photosynthesis), the bathypelagic (1,000 - 4,000 meters), the abyssopelagic (4,000 - 6,000 meters) and the deepest, the hadopelagic (the deep trenches below 6,000 meters to about 11,000 m or 36,000 feet deep). The last three zones have no sunlight at all.

Benthic zones are defined as the bottom sediments and other surfaces of a body of water such as an ocean or a lake. Organisms living in this zone are called benthos. They live in a close relationship with the bottom of the sea, with many of them permanently attached to it, some burrowed in it, others swimming just above it. In oceanic environments, benthic habitats are zoned by depth, generally corresponding to the comparable pelagic zones: the intertidal (where sea meets land, with no pelagic equivalent), the subtidal (the continental shelves, to about 200 m), the bathyal (generally the continental slopes to 4,000 m), the abyssal (most of the deep ocean seafloor, 4,000 - 6,000 m), and the hadal (the deep trenches 6,000 to 11,000 m).

There are several types of deep benthic surfaces, each having different life forms. First, most of the deep seafloor consists of mud (very fine sediment particles) or "ooze" (defined as mud with a high percentage of organic remains) due to the accumulation of pelagic organisms that sink after they die. [Unlike the shoreline, sandy habitats are rarely found in the deep sea because sand particles, created by wave action on coral and rocks at shorelines, are too heavy to be carried by currents to the deep.] Second, benthic areas too steep for sediment to stick are rocky. Rocky areas are found on the flanks of islands, seamounts, rocky banks, on mid-ocean ridges and their rift valleys, and some parts of continental slopes. At the mid-ocean ridges, where magma wells up and pushes sea-floor tectonic plates apart, even flat surfaces are rocky because these areas are too geologically new to have accumulated much mud or ooze. Third, in some areas certain chemical reactions produce unique benthic formations. The best known of these formations are the "smoker" chimneys created by hydrothermal vents.

Exploration of these zones has presented a challenge to scientists for decades and much remains to be discovered. However, advances in technology are increasingly allowing scientists to learn more about the strange and mysterious life that exists in this harsh environment. Life in the deep sea must withstand total darkness (except for non-solar light such as bioluminescence), extreme cold, and great pressure. To learn more about deep-sea marine life, sophisticated data collection devices have been developed to collect observations and even geological and biological samples from the deep. First, advances in observational equipment such as fiber optics that use LED light and low light cameras has increased our understanding of the behaviors and characteristics of deep sea creatures in their natural habitat. Such equipment may be deployed on permanent subsea stations connected to land by fiber optic cables, or on "lander" devices which drop to the seafloor and which are later retrieved (typically after a radio command activates the dropping of ballast so the lander may float up.) Second, remotely operated vehicles (ROVs) have been used underwater since the 1950s. ROVs are basically unmanned submarine robots with umbilical cables used to transmit data between the vehicle and researcher for remote operation in areas where diving is constrained by physical hazards. ROVs are often fitted with video and still cameras as well as with mechanical tools such as mechanical arms for specimen retrieval and measurements. Other unmanned submarine robots include AUVs (autonomous undersea vehicles) that operate without a cable, and the USA's new Nereus, a hybrid unmanned sub which can switch from ROV to AUV mode and which is currently the world's only unmanned submarine capable of reaching the deepest trenches. Third, manned deep sea submersibles are also used to explore the ocean's depths. Alvin is an American deep sea submersible built in 1964 that has been used extensively over the past 4 decades to shed light on the black ocean depths. Like ROVs, it has cameras and mechanical arms. This sub, which carries 3 people (typically a pilot and 2 scientists), has been used for more than

4,000 dives W a maximum depth of more than 4,500 m. France, Japan and Russia have similar manned scientific submersibles that can reach somewhat greater depths, while China is currently building one to reach 7,000 m.

The bathyscaphe Trieste at the National Museum of the U.S. Navy in Washington, D.C.

Until 2012, only one manned submarine device has ever reached the bottom of Mariana trench at almost 11,000 m: the bathyscaphe Trieste manned by Jacques Piccard and Don Walsh. During the Trieste's single dive in 1960, its windows began to crack, and it has never been used since. 52 years later, on March 25, 2012 (March 26 local time), James Cameron successfully dove in his commissioned one-man sub to the Challenger Deep. Don Walsh was invited to join the expedition.

Physical Characteristics of the Deep Sea

The physical characteristics that deep sea life must contend with to survive are:

1. abiotic (non-living) ones, namely light (or lack thereof), pressure, currents, temperature, oxygen, nutrients and other chemicals; and

2. biotic ones, that is, other organisms that may be potential predators, food, mates, competitors or symbionts.

All these factors have led to fascinating adaptions of deep sea life for sensing, feeding, reproducing, moving, and avoiding being eaten by predators.

Light

The deep sea begins below about 200 m, where sunlight becomes inadequate for photosynthesis. From there to about 1,000 m, the mesopelagic or "twilight" zone, sunlight continues to decrease until it is gone altogether. This faint light is deep blue in color because all the other colors of light are absorbed at depth. The deepest ocean waters below 1,000 m are as black as night as far as sunlight is concerned. And yet, there is some light. People who dive deep in a submersible (with its lights off) are often mesmerized by an incredible "light show" of floating, swirling, zooming flashes of light. This is bioluminescence, a chemical reaction in a microbe or animal body that creates light without heat, and it is very common. And yet, this light is low compared to sunlight, so animals here — as well as those in the mesopelagic zone — need special sensory adaptations. Many deep-sea fish such as the stout blacksmelt have very large eyes to capture what little light exists. Other animals such as tripodfishes are essentially blind and instead rely on other, enhanced senses including smell, touch and vibration.

Scientists think bioluminescence has six different functions (not all used by any one species):

1. Headlights, such as the forward-facing light organs (called photophores) of lantern fish;

2. Social signals such as unique light patterns for attracting mates;

3. Lures to attract curious prey, such as the dangling "fishing lures" of anglerfish;

4. Counterillumination, in which rows of photophores on the bellies of many mesopelagic fish produce blue light exactly matching the faint sunlight from above (making the fish invisible to predators below them);

5. Confusing predators or prey, such as bright flashes that some squid make to stun their prey, and decoys that divert attention, such as the glowing green blobs ejected by green bomber worms; and

6. "Burglar alarms" in which an animal being attacked illuminates its attacker (the "burglar") so that an even bigger predator (the "police") will see the burglar and go after it. Some swimming sea cucumbers even coat their attackers with sticky glowing mucus so the "police" predators can find them many minutes later.

Most bioluminescence is blue, or blue-green, because those are the colors that travel farthest in water. As a result, most animals have lost the ability to see red light, since that is the color of sunlight that disappears first with depth. But a few creatures, like the dragonfish, have evolved the ability to produce red light. This light, which the dragonfish can see, gives it a secret "sniper" light to shine on prey that do not even know they are being lit up.

Pressure

Considering the volume of water above the deepest parts of the ocean, it's no wonder that hydrostatic pressure is one of the most important environmental factors affecting deep sea life. Pressure increases 1 atmosphere (atm) for each 10 m in depth. The deep sea varies in depth from 200 m to about 11,000 m, therefore pressure ranges from 20 atm to more than 1,100 atm. High pressures can cause air pockets, such as in fish swim bladders, to be crushed, but it does not compress water itself very much. Instead, high pressure distorts complex biomolecules — especially membranes and proteins — upon which all life depends. Indeed, many food companies now use high pressure to sterilize their products such as packaged meats.

Life appears to cope with pressure effects on biomolecules in two ways. First, their membranes and proteins have pressure-resistant structures that work by mechanisms not yet fully understood, but which also mean their biomolecules do not work well under low pressure in shallow waters. Second, some organisms may use "piezolytes". These are small organic molecules recently discovered that somehow prevent pressure from distorting large biomolecules. One of these piezolytes is trimethylamine oxide (TMAO). This molecule is familiar to most people because it gives rise to the fishy smell of marine fish and shrimp. TMAO is found at low levels in shallow marine fish and shrimp that humans routinely eat, but TMAO levels increase linearly with depth and pressure in other species. Really deep fish, including some grenadiers which humans are now fishing, smell much more fishy.

Animals brought from great depth to the surface in nets and submersible sample boxes generally die; in the case of some (but not most) deep-sea fishes, their gas-filled swim bladder (adapted to resist high pressure) expands to a deadly size. However, the vast majority of deep-sea life has no air pockets that would expand as pressure drops during retrieval. Instead, it is thought that rapid pressure as well as temperature changes kill them because their biomolecules no longer work well (high TMAO does not help, as it appears to be too high in deep-sea life for biomolecules to work properly at the surface). Advances in deep sea technology are now enabling scientists to collect species samples in chambers under pressure so that they reach the surface for study in good condition.

Pressure-adapted microbes have been retrieved from trenches down to 11,000 m, and have been found in the laboratory to have all these adaptations (pressure-resistant biomolecules and piezolytes). However, pressure adaptations have only been studied in animals down to about 5,000 m. We do not yet know if the adaptations found at those depths work at greater depths down to 11,000 m.

Temperature

Except in polar waters, the difference in temperature between the euphotic, or sunlit, zone near the surface and the deep sea can be dramatic because of thermoclines, or the separation of water layers of differing temperatures. In the tropics, for example, a layer of warm water over 20°C floats on top of the cold, dense deeper water. In most parts of the deep sea, the water temperature is more uniform and constant. With the exception of hydrothermal vent communities where hot water is emitted into the cold waters, the deep sea temperature remains between about -1 to about +4°C. However, water never freezes in the deep sea (note that, because of salt, seawater freezes at -1.8°C). If it did somehow freeze, it would just float to the surface as ice. Life in the deep is thought to adapt to this intense cold in the same ways that shallow marine life does in the polar seas. This is by having "loose" flexible proteins and unsaturated membranes which do not stiffen up in the cold. Membranes are made of fats and need to be somewhat flexible to work well, so you may be familiar with this adaptation in your kitchen. Butter, a saturated fat, is very hard in your refrigerator and would make a poor membrane in the cold, while olive oil — an unsaturated fat — is semi-solid and would make a good flexible membrane. However, as with pressure, there is a tradeoff: loose membranes and proteins of cold-adapted organisms readily fall apart at higher temperatures (much as olive oil turns to liquid at room temperature).

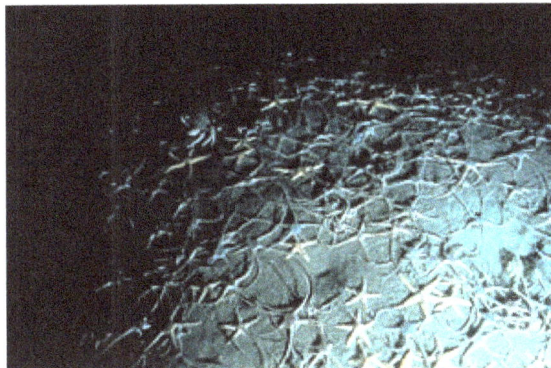

Oxygen

The dark, cold waters of much of the deep sea have adequate oxygen. This is because cold water can dissolve more oxygen than warm water, and the deepest waters generally originate from shallow polar seas. In certain places in the northern and southern seas, oxygen-rich waters cool off so much that they become dense enough to sink to the bottom of the sea. These so-called thermohaline currents can travel at depth around the globe, and oxygen remains sufficient for life because there is not enough biomass to use it all up. However, there are also oxygen-poor environments in intermediate zones, wherever there is no oxygen made by photosynthesis and there are no thermohaline currents. These areas, called oxygen minimum zones, usually lie at depths between 500 - 1,000 m in temperate and tropical regions. Here, animals as well as bacteria that feed on decaying food particles descending through the water column use oxygen, which can consequently drop to near zero in some areas. Biologists are still investigating how animals survive under such conditions.

Although most of the deep seafloor has oxygen, there are exceptions in isolated basins with no circulation. Some of these basins that have no oxygen are found at the bottom of the Mediterranean

Sea. In 2010, scientists investigating these at 3,000 m depths made a startling discovery: the first known animals to be living continuously without any oxygen. The animals are tiny Loriciferans, members of an animal phylum first discovered in 1983. How they can survive these conditions is not yet known.

Food

Deep sea creatures have evolved some fascinating feeding mechanisms because food is scarce in these zones. In the absence of photosynthesis, most food consists of detritus — the decaying remains of microbes, algae, plants and animals from the upper zones of the ocean — and other organisms in the deep. Scavengers on the seafloor that eat this "rain" of detritus include sea cucumbers (the most common benthic animal of the deep), brittle stars and grenadier or rattail fish. The corpses of large animals such as whales that sink to the bottom provide infrequent but enormous feasts for deep sea animals and are consumed by a variety of species. This includes jawless fish such as hagfish, which burrow into carcasses, quickly consuming them from the inside out; scavenger sharks; crabs; and a newly discovered group of worms (called Osedax, meaning bone-eater) which grow root-like structures into the bone marrow.

Deep-sea pelagic fish such as gulper eels have very large mouths, huge hinged jaws and large and expandable stomachs to engulf and process large quantities of scarce food. Many deep-sea pelagic fish have extremely long fang-like teeth that point inward. This ensures that any prey captured has little chance of escape. Some species, such as the deep sea anglerfish and the viperfish, are also equipped with a long, thin modified dorsal fin on their heads tipped with a photophore lit with bioluminescence used to lure prey. Many of these fish don't expend much energy swimming in search of food; rather they remain in one place and ambush their prey using clever adaptations such as these lures. Others, such as rattails or grenadiers cruise slowly over the seafloor listening and smelling for food sources failing from above, which they engulf with their large mouths.

Many mesopelagic and deeper pelagic species also save energy by having watery, gelatinous muscles and other tissues with low nutritive content. For example, an epipelagic tuna's muscle (the kind you might eat) may be 20% protein. This makes for a strong, fast muscle, but also takes considerable energy to maintain. In contrast, a deep pelagic blacksmelt or viperfish may have only 5-8% protein. This means they cannot swim as well as a tuna, but they can achieve a larger body size with much less maintenance costs.

Some mesopelagic species have adapted to the low food supply (and sometimes to the low oxygen content) in moderate-depth waters with a special behavior called vertical migration. At dusk, millions of lantern fish, shrimp, jellies and other mobile animals migrate to the food-rich surface waters to feed in the darkness of night. Then, presumably to avoid being eaten in daylight, they return to the depths at dawn to digest. Some of the species undergo large pressure and temperature changes during their daily migrations, but we do not yet know exactly how they cope with those dramatic daily changes.

Since plankton is scarce in the deep sea, filter feeding (the most common mode of feeding in shallow waters) is a difficult way to make a living. Consequently, some deep-sea animals belonging to groups once thought to be exclusively filter feeders have evolved into carnivores. One of these is the carnivorous sea squirt Megalodicopia hians. Sea squirts or tunicates are generally harmless filter feeders which draw in microscopic organisms through a siphon tube, but Megalodicopia hians has a huge jaw-like siphon that can rapidly engulf swimming animals. Another of these is the ping-pong-tree sponge, Chondrocladia lampadiglobus. Again, the vast majority of sponges draw in microscopic material through tiny pores, but this sponge has tree-like branches with large glass globes covered in Velcro-like sharp spikes that impale swimming prey.

Other Adaptations of Deep-sea Animals

Deep-sea animals have evolved many unique adaptations to cope with their harsh environment. Let's look at some others, not all of which are fully understood.

- Body Color: This is often used by animals everywhere for camouflage and protection from predators. In the deep sea, animals' bodies are often transparent (such as many jellies and squids), black (such as blacksmelt fish), or even red (such as many shrimp and other squids). The absence of red light at these depths keeps them concealed from both predators and prey. Some mesopelagic fish such as hatchetfish have silvery sides that reflect the faint sunlight, making them hard to see.

- Reproduction: Consider how hard it must be to find a mate in the vast dark depths. For most deep sea species, we do not know how they achieve this. Earlier we noted that unique light patterns may aid in this. Deep-sea anglerfish may use such light patterns as well as scents to find mates, but they also have another interesting reproductive adaptation. Males are tiny in comparison to females and attach themselves to their mate using hooked teeth, establishing a parasitic-like relationship for life. The blood vessels of the male merges with the female's so that he receives nourishment from her. In exchange, the female is provided with a very reliable sperm source, avoiding the problem of having to locate a new mate every breeding cycle.

- Gigantism: Another possible adaptation that is not fully understood is called deep-sea gigantism. This is the tendency for certain types of animals to become truly enormous in size. A well-known example is the giant squid, but there are many others such as the colossal squid, the giant isopod, the king-of-herrings oarfish (which may be the source of sea-serpent legends), and the recently captured giant amphipod from 7,000 m in the Kermadec Trench near New Zealand. While the giant tubeworms of hydrothermal vents grow well due to abundant energy supplies, the other gigantic animals live in food-poor habitats, and it is not known how they achieve such growth. It may simply be a result of the feature-long lives.

- Long Lives: Many deep-sea organisms, including gigantic but also many smaller ones, have been found to live for decades or even centuries. Long-lived fishes include rattails or grenadiers and the orange roughy, which are of special concern as they are targets of deep-sea fisheries. These species reproduce and grow to maturity very slowly, such that populations may take decades to recover (if at all) after being overfished. This has happened repeatedly to the orange roughy, a deep-sea fish easily found congregating around seamounts in the southern oceans. Once fisheries have wiped out one seamount population, they move on to another seamount.

Also of concern with respect to their long, slow lives are a group of animals once thought to be restricted to warm tropical waters: corals. In the last 30 years, numerous cold-water coral species have been found on rocky surfaces throughout the deep sea. These animal colonies may live for centuries, or — amazingly — even millennia. One deep-sea coral colony off Hawaii has been dated at over 4,000 years old, making it older than the Pyramids of Egypt. Again, these corals are highly vulnerable to fisheries as they are easily destroyed by deep-sea trawl nets, and they may take decades to grow back.

Hydrothermal Vent and Cold Seep Communities

Life in the deep sea is relatively sparse compared to the epipelagic (euphotic) and intertidal zones, with two exciting, and relatively recently discovered exceptions — hydrothermal vent and cold seep communities.

Hydrothermal Vents

These amazing formations were first discovered in 1976 - 77 during a deep sea expedition with Alvin at a mid-ocean ridge near the Galapagos. At that time, only geologists were aboard, with the goal of directly observing seafloor spreading — the mid-ocean ridges being places where magma welling up underneath pushes two tectonic plates apart, creating a rift valley between them. Some geologists thought there might be geyser-like hot springs, as found in rift valleys on land (such as in Iceland), while others thought that high pressure would prevent such formations. However no one predicted any interesting biology. What they found not only revolutionized geology but biology even more so. These dives to depths of about 2,700 m revealed hot springs of far greater complexity and beauty than anyone had imagined: hot mineral-rich water spewing (like continuous geysers) from vents heated by magma, with metal sulfides precipitating in the cold surrounding seawater to form intricate, colorful and often towering chimneys.

Moreover, a completely unexpected community of life was found around these aptly named hydrothermal vents, with not only high densities of numerous new species, but also a new kind of ecosystem flourishing in the dark that had never been imagined by scientists — an ecosystem based on toxic gas. The most amazing of the new species was a giant tubeworm, named Riftia. Growing rapidly in dense clusters, these 2-meter-tall worms were found to have no digestive tract. The revolutionary finding was that they subsist on energy-rich hydrogen sulfide in vent water and generated in the Earth's crust. Hydrogen sulfide (rotten-egg gas) is normally toxic to animals, but these worms avoid the problem in a spectacular manner. They harbor bacteria known as chemoautotrophs (in a large sac replacing a digestive system), which can use the energy in hydrogen sulfide to convert carbon dioxide into sugars, just as plants do using sunlight. The worm's blood picks up and delivers sulfide, carbon dioxide and oxygen to these bacterial symbionts, which in turn "feed" their hosts with the excess sugars they make (while turning the sulfide into a non-toxic waste product). Thus, the ecosystem was found to run on the Earth's geothermal energy rather than sunlight. Many scientists now think that life on Earth began at such vents over 3 billion years ago.

Since those first discoveries near the Galapagos, hydrothermal vent communities have been found at depths ranging from about 1,500 m to over 5,000 m. Most vents are along the mid-ocean ridges, where magma is close to seawater. Other animals with bacterial symbionts have been found, including other species of tubeworms, giant clams and mussels, snails, and shrimp. Undoubtedly many vent communities are yet to be found, since many ridge areas have not yet been explored.

The water temperature of vents is much warmer than normal for the deep sea (about 2°C), reaching as high at 400°C without boiling due to the high pressure. However, nothing can live at such temperatures. The communities of vent life are mostly found between about 8 - 25°C, but may reach perhaps 60°C around some animals such as Pompeii worms (Alvinella). Though complex life seems not to live at higher temperatures, some archaea have been found living at temperatures of over 120°C.

Cold Seeps

After the first vent discoveries, other unexpected high-density deep-sea ecosystems were found. Named cold seeps, these occur at places (mostly along continental margins) where cold methane (which at depths below 500 m forms methane-hydrate "ice"), hydrogen sulfide, and/or oil seep out of sediments to provide abundant energy. By some estimates, there is more energy locked up in the methane hydrates than in all (other) fossil/hydrocarbon fuels combined. Animals with symbiotic bacteria were found, different from but related to vent species, including tubeworms, clams, and mussels. Some mussels harbor methane-using bacteria instead of sulfide-using ones, making ecosystems powered by natural gas.

Dense seep communities have also been found around deep brine pools, or "lakes within oceans." These form where salt deposits under the ocean floor dissolve to form pools of water so dense from their salt content that they do not mix with the overlying seawater. So far a few of these have been found in the Gulf of Mexico and the Mediterranean Sea. At the best-studied brine pool (in the Gulf of Mexico: high densities of mussels live around the rim, subsisting (using symbionts) on methane gas seeping from the pool. However, no known animal can survive the salt within the pool itself. Various microbes have been found in the high salt waters, however.

In 2012, a new deep-sea ecosystem dubbed a "hydrothermal seep" was discovered off Costa Rica. It is a mosaic of vent and seep communities, with many new species.

References

- Vásquez, J.A., J.M. Alonso Vega and A.H. Buschmann. 2006. Long term variability in the structure of kelp communities in northern Chile and the 1997-98 ENSO. Journal of Applied Phycology 18: 505-519

- Aquatic-microbiology, boundless-microbiology: lumenlearning.com, Retrieved 24 April 2018

- Kelp forests provide habitat for a variety of invertebrates, fish, marine mammals, and birds NOAA. Updated 11 January 2013. Retrieved 15 January 2014

- What-is-a-mangrove-2291773: thoughtco.com, Retrieved 10 June 2018

- Seagrass-and-seagrass-beds, plants-algae: ocean.si.edu, Retrieved 25 May 2018

- Duggins, D.O., J.E. Eckman and A.T. Sewell. 1990. Ecology of understory kelp environments. II. Effects of kelps on recruitment of benthic invertebrates. Journal of Experimental Marine Biology and Ecology 143: 27-45

- Open-ocean-pelagic-zone-2291774: thoughtco.com, Retrieved 16 March 2018

- Surface-water-2360: safeopedia.com, Retrieved 19 July 2018

Chapter 4

Marine Ecology

Marine ecosystems are large aquatic ecosystems that include salt marshes, lagoons, estuaries, coral reefs, the deep sea, etc. The study of marine ecosystems is within the scope of marine ecology. All the diverse aspects of marine ecology including aquatic and fresh water ecology have been carefully analyzed in this chapter.

Marine ecology is an integrative science that studies the basic structural and functional relationships within and among living populations and their physical-chemical environments in marine ecosystems. Marine ecology draws on all the major fields within the biological sciences as well as oceanography, physics, geology, and chemistry. Emphasis has evolved toward understanding the rates and controls on ecological processes that govern short- and long-term events, including population growth and survival, primary and secondary productivity, and community dynamics and stability. Marine ecology focuses on specific organisms as well as on particular environments or physical settings.

Marine Environments

Classification of marine environments for ecological purposes is based very generally on two criteria, the dominant community or ecosystem type and the physical-geological setting. Those ecosystems identified by their dominant community type include mangrove forests, coastal salt marshes, submersed seagrasses and seaweeds, and tropical coral reefs. Marine environments identified by their physical-geological setting include estuaries, coastal marine and nearshore zones, and open-ocean-deep-sea regions.

An estuary is a semi enclosed area or basin with an open outlet to the sea where fresh water from the land mixes with seawater. The ecological consequences of fresh-water input and mixing create strong gradients in physical-chemical characteristics, biological activity and diversity, and the potential for major adverse impacts associated with human activities. Because of the physical forces of tides, wind, waves, and fresh-water input, estuaries are perhaps the most ecologically complex marine environment. They are also the most productive of all marine ecosystems on an area basis and contain within their physical boundaries many of the principal marine ecosystems defined by community type.

Coastal and near shore marine ecosystems are generally considered to be marine environments bounded by the coastal land margin (seashore) and the continental shelf 300–600 ft (100– 200 m) below sea level. The continental shelf, which occupies the greater area of the two and varies in width from a few to several hundred kilometers, is strongly influenced by physical oceanographic processes that govern general patterns of circulation and the energy associated with waves and currents. Ecologically, the coastal and nearshore zones grade from shallow water depths, influenced by the adjacent landmass and input from coastal rivers and estuaries, to the continental shelf break, where oceanic processes predominate. Biological productivity and species diversity and abundance tend to decrease in an offshore direction as the food web becomes supported only by planktonic production. Among the unique marine ecosystems associated with coastal and near shore water bodies are seaweed-dominated communities (for example, kelp "forests"), coral reefs, and upwellings.

Approximately 70% of the Earth's surface is covered by oceans, and more than 80% of the ocean's surface overlies water depths greater than 600 ft (200 m), making open-ocean–deep-sea environments the largest, yet the least ecologically studied and understood, of all marine environments. The major oceans of the world differ in their extent of landmass influence, circulation patterns, and other physical-chemical properties. Other major water bodies included in open-ocean–deep-sea environments are the areas of the oceans that are referred to as seas. A sea is a water body that is smaller than an ocean and has unique physical oceanographic features defined by basin morphology. Because of their circulation patterns and geomorphology, seas are more strongly influenced by the continental landmass and island chain structures than are oceanic environments.

Within the major oceans, as well as seas, various oceanographic environments can be defined. A simple classification would include water column depths receiving sufficient light to support photosynthesis (photic zone); water depths at which light penetration cannot support photosynthesis and which for all ecological purposes are without light (aphotic zone); and the benthos or bottom-dwelling organisms. Classical oceanography defines four depth zones; epipelagic, 0–450 ft (0–150 m), which is variable; mesopelagic, 450–3000 ft (150–1000 m); bathypelagic, 3000–12,000 ft (1000–4000 m); and abyssopelagic, greater than 12,000 ft (4000 m). These depth strata correspond approximately to the depth of sufficient light penetration to support photosynthesis; the zone in which all light is attenuated; the truly aphotic zone; and the deepest oceanic environments.

Marine Ecological Processes

Fundamental to marine ecology is the discovery and understanding of the principles that underlie the organization of marine communities and govern their behavior, such as controls on population growth and stability, quantifying interactions among populations that lead to persistent communities, and coupling of communities to form viable ecosystems. The basis of this organization is the flow of energy and cycling of materials, beginning with the capture of radiant solar energy through the processes of photosynthesis and ending with the remineralization of organic matter and nutrients.

Photosynthesis in seawater is carried out by various marine organisms that range in size from the microscopic, single-celled marine algae to multicellular vascular plants. The rate of photosynthesis, and thus the growth and primary production of marine plants, is dependent on a number of factors, the more important of which are availability and uptake of nutrients, temperature, and intensity and quality of light. Of these three, the last probably is the single most important in governing primary production and the distribution and abundance of marine plants. Considering the high attenuation of light in water and the relationships between light intensity and photosynthesis, net autotrophic production is confined to relatively shallow water depths. The major primary producers in marine environments are intertidal salt marshes and mangroves, submersed seagrasses and seaweeds, phytoplankton, benthic and attached microalgae, and—for coral reefs—symbiotic algae (zooxanthellae). On an areal basis, estuaries and near shore marine ecosystems have the highest annual rates of primary production. From a global perspective, the open oceans are the greatest contributors to total marine primary production because of their overwhelming size.

The two other principal factors that influence photosynthesis and primary production are temperature and nutrient supply. Temperature affects the rate of metabolic reactions, and marine plants show specific optima and tolerance ranges relative to photosynthesis. Nutrients, particularly nitrogen, phosphorus, and silica, are essential for marine plants and influence both the rate of photosynthesis and plant growth. For many phytoplankton-based marine ecosystems, dissolved inorganic nitrogen is considered the principal limiting nutrient for autotrophic production, both in its limiting behavior and in its role in the eutrophication of estuarine and coastal waters.

Marine food webs and the processes leading to secondary production of marine populations can be divided into plankton-based and detritus-based food webs. They approximate phytoplankton-based systems and macrophyte-based systems. For planktonic food webs, current evidence suggests that primary production is partitioned among groups of variously sized organisms, with small organisms, such as cyanobacteria, playing an equal if not dominant role at times in aquatic productivity. The smaller autotrophs—both through excretion of dissolved organic compounds to provide a substrate for bacterial growth and by direct grazing by protozoa (microflagellates and ciliates)—create a microbially based food web in aquatic ecosystems, the major portion of autotrophic production and secondary utilization in marine food webs may be controlled, not by the larger organisms typically described as supporting marine food webs, but by microscopic populations.

Macrophyte-based food webs, such as those associated with salt marsh, mangrove, and seagrass ecosystems, are not supported by direct grazing of the dominant vascular plant but by the production of detrital matter through plant mortality. The classic example is the detritus-based food webs of coastal salt marsh ecosystems. These ecosystems, which have very high rates of primary production, enter the marine food web as decomposed and fragmented particulate organics. The particulate organics of vascular plant origin support a diverse microbial community that includes bacteria, flagellates, ciliates, and other protozoa. These organisms in turn support higher-level consumers.

Both pelagic (water column) and benthic food webs in deep ocean environments depend on primary production in the overlying water column. For benthic communities, organic matter must reach the bottom by sinking through a deep water column, a process that further reduces its energy content. Thus, in the open ocean, high rates of secondary production, such as fish yields, are associated with areas in which physical-chemical conditions permit and sustain high rates of primary production over long periods of time, as is found in upwelling regions.

Regardless of specific marine environment, microbial processes provide fundamental links in marine food webs that directly or indirectly govern flows of organic matter and nutrients that in turn control ecosystem productivity and stability.

The vastness of the oceans, the largest continuous environment on earth, has provided a safe shelter for about 20% of all living organisms until the beginning of industrial revolution. Since then, this once invincible environment has been under constant change and destruction, the results of which now are threatening all life forms on earth. With this rate of destruction, we are possibly losing our window of opportunity to protect aquatic biodiversity and learn how aquatic organisms evolved to find ways for adapting life in water. Marine Ecology, in its simplest terms the study of marine organisms and their habitats, continues to provide fundamental information to better understand the effects of global changes on eco-biology of organisms.

In many cases, marine ecology is more intricate than the relatively simple study of a specific living organism or its environment due to various intra and inter specific interactions between other organisms and due to effects of numerous factors on a particular environment. Therefore, marine ecologists rather than concentrating on a single species, organism or habitat, often find themselves simultaneously focusing on interactions between organisms and the effects of environmental factors on these organisms. During the last decades, the complex nature of these interactions is being exacerbated due to the changes induced by a variety of factors such as increased ocean temperatures, dramatic changes in weather patterns, ocean acidification, melting of glaciers, and pollution. The effects of these man-made factors are occurring in a relatively shorter time scale and in many cases are beyond the capacity of organisms to adapt to these deviations. Throughout the world, new conditions are often manifesting themselves as loss of biodiversity accompanied with other major changes such as shifts in distributions of many species toward higher latitudes and changes in timing of one of the most important factors that influence life in the oceans is temperature. Temperature affects the rate biological processes proceed. In general, the metabolic rate of poikilotherms doubles with a 10°C increase in temperature. However, much less temperature differences are enough to trigger changes in weather patterns that have worldwide effects. Global mean temperatures are now 0.50°C higher than it was since 1960s. A typical example of increased global temperatures is the El Niño phenomenon that occurs periodically over the Pacific Ocean and characterized by increased temperatures of surface waters. It is well established that increased water temperatures results in weakened currents and less rain in the Southern Ocean, which in turn, results in dramatic changes in physicochemical and biological conditions. Fluctuation in nutrient concentrations is the most notable factor that altered circulation pattern effects. Such interruptions of nutrient fluxes have important consequences on the primary production which in turn affect fish stocks. The relationship between fluctuations in the abundance of anchovy in the Southeast Pacific Ocean and the periodicity of El Niño has been established. The fluctuations in the abundance of these commercial fish stocks have important socioeconomic consequences due to enormous yields, which fluctuated between 3 and 8 million tons during the last decade.

Another important parameter that influences life in the ocean is CO_2 levels in the atmosphere. As a result of global industrialization, CO_2 levels have increased over the last 100 years. Higher CO_2 levels in the atmosphere forces this gas into the surface waters which results in lower pH values. As a result, the mean pH value of the earth's oceans has fallen by 0.10 pH units . Insignificant as it may seem, this drop corresponds roughly to 30% increase in the concentration of hydrogen ions. Organisms such as corals, bivalves, and calcareous plankton are susceptible to reduced pH levels as acidic conditions dissolve calcium carbonate. Therefore, the disruption of the calcification process may have serious consequences due to its potential to negatively affected calcareous species in the food web.

Another important factor that is becoming increasingly influential on all life on earth is the increasing rate of melting of ice in Polar Regions. The melting of ice causes a series of events including, sea level rise, freshening of seawater, and reduction in the speeds of major current systems in the oceans. While sea level rise will have catastrophic effects mainly for human habitation in coastal areas, freshening of seawater and its effect on currents will potentially affect all life forms due to the changes in global climate.

Although the effects of individual stressors are relatively well studied, there are limited data on compounded effects of multiple stressors. Stressors such as temperature, salinity, UV, hypoxia, acidification, and pollution may be simultaneously experienced by marine organisms, especially in coastal areas. In many cases, organisms exposed to multiple stressors exhibit reduced resistance. For example, many coral reefs are simultaneously suffering from increasing temperatures, acidification, diseases, and silting. Toxicity of pollutants has been shown to increase salinity or temperature stress. This is particularly important because even if strict fisheries regulations become effective for a particular overexploited area in a heavily modified coastal system, expected recovery of stocks may not be possible due to increased vulnerability of early life stages to multiple stressors relative to juvenile or adult stages.

While it is relatively easier to observe the effects of altered physicochemical conditions over larger scales, the effects of pollution and over exploitation are relatively easier to observe in smaller scales. A typical example is the Black Sea which is closed basins with limited water exchange rates and relatively smaller surface areas. Between the period 1950 and 1970, the Turkish Black Sea fishery was characterized by larger predators such as tuna, swordfish, and bonito. Following a decrease in top predators as a result of increased fishing pressure, industrial fishing operations concentrated on small pelagic fish species such as anchovy and sardine. Therefore, after 1970s, there was a major shift in commercial fishing operations. Due to the developments in industrial fishing methods, a steady increase was observed until late 1980s with a maximum of 600,000 tons in 1988. This increase in fish productivity was correlated with a 10-fold increase in phytoplankton biomass in the 1980s compared to that of 2–3 g m−2 in 1960s. This dramatic increase was due to increased inputs of agricultural nitrates and phosphates into the Black Sea through rivers and the subsequent mixing of these nutrients in the water column. As a result of this enrichment, primary production was able to support—despite increased fishing pressure—high yields of small pelagic fishes for almost a decade before a major collapse observed after late 1980s. For example, in 1990, anchovy landings were only 66,000 tons, which was less than ¼ of that in 1988. This collapse in small pelagic fisheries was also experienced by other nations bordering the Black Sea and as a result, total landings dropped down to 200 thousand tons in 1991, compared to that of 900 thousand tons in 1988. It is believed that overexploitation was not the only factor for the simultaneous collapse in small pelagic fish stocks experienced throughout the Black Sea. The lobate ctenophore, Mnemiopsis leidyi A. Agassiz which was reported for the first time in the Black Sea in 1982, had reached a biomass of up to 1 kg m−3 by the end of 1980s. Its broad tolerance to a variety of physicochemical conditions, rapid growth and voracious appetite for zooplankton, fish eggs, and larvae has contributed significantly to the collapse of Black Sea fisheries. After 1990s, although reductions in concentrations of agricultural nutrients in the Black Sea and the introduction of Beroe ovata Mayer, 1912, the pink comb jellyfish that feeds on M. leidyi tipped the balance in favor of recovery of small pelagic fish stocks, we are still miles away from the point of sustainable management of fisheries in the Black Sea. Yet, even over a relatively smaller scale and with no diverse multinational management strategies that can limit the success of management programs, fisheries in the Sea of Marmara is almost an identical episode of what was experienced in the Black Sea. For example, a comparison of catch rates in 1990 and 2015 showed a 1.50- to 130-fold decrease in all reported demersal species as a result of eutrophic conditions as indicated by increased periodicity and intensity of phytoplankton blooms , introduction of M. leidyi in early 1990s , continuous heavy fishing pressure and lack of effective management strategies. Recovery efforts for these two interconnected ecosystems will require a multidisciplinary approach to rebuild fishery resources. Unfortunately, decreasing fish stocks is not only an issue of semi-closed basins with highly populated areas. It is estimated that globally up to 63% of fish stocks are in need of rebuilding and efforts toward rebuilding diversity will meet major challenges considering human-induced and global-scale impacts.

Aquatic

Aquatic ecosystems are any water-based environment in which plants and animals interact with the chemical and physical features of the aquatic environment. Aquatic ecosystems are generally divided into two types --the marine ecosystem and the freshwater ecosystem. The largest water ecosystem is the marine ecosystem, covering over 70 percent of the earth's surface. Oceans, estuaries, coral reefs and coastal ecosystems are the various kinds of marine ecosystems. Freshwater ecosystems cover less than 1 percent of the earth and are subdivided into lotic, lentic and wetlands.

Ocean Ecosystems

The earth has five major oceans: Pacific Ocean, Indian Ocean, Arctic Ocean, Atlantic Ocean and Southern (Antarctic) Ocean. Although the oceans are connected, each of them has unique species and features. The Pacific is the largest and deepest ocean and the Atlantic is second in size.

Oceans are home to different species of life. The waters of the Arctic and Southern Oceans are very cold, yet filled with life. The largest population of krill (small, shrimp-like marine creatures) lies under the ice of the Southern Ocean.

Life in Estuaries

Estuaries are places where rivers meet the sea and may be defined as areas where salt water is diluted with fresh water. River mouths, coastal bays, tidal marshes and water bodies behind barrier beaches are some examples of estuaries. They are biologically productive as they have a special kind of water circulation that traps plant nutrients and stimulates primary production.

Coral Reefs

According to Environmental Protection Agency, coral reefs are the world's second richest ecosystems and have a wide diversity of plants and animals. As a result, coral reefs often are referred to as the rain forest of the oceans.

Coastal Systems

Land and water join to create the coastal ecosystems. These ecosystems have a distinct structure, diversity, and flow of energy. Plants and algae are found at the bottom of the coastal ecosystem. The fauna is diverse and consists of insects, snails, fish, crabs, shrimp, lobsters etc.

Lotic Ecosystems

Lotic ecosystems are the systems with rapid flowing waters that move in a unidirectional way such as rivers and streams. These environments harbor numerous species of insects such as mayflies, stoneflies and beetles which have developed adapted features such as weighted cases to survive the environment. Several species of fishes such as eel, trout and minnow are found here. Various mammals such as beavers, otters and river dolphins inhabit lotic ecosystems.

Lentic Ecosystems

Lentic ecosystems include all standing water habitats such as lakes and ponds. These ecosystems are home to algae, rooted and floating-leaved plants and invertebrates such as crabs and shrimps. Amphibians such as frogs and salamanders and reptiles like alligators and water snakes are also found here.

Swamps and Wetlands

Wetlands are marshy areas and are sometimes covered in water, which have a wide diversity of plants and animals. Swamps, marshes, and bogs are some examples in this regard. Plants such as black spruce and water lilies are commonly found in wetlands. The fauna consists of dragonflies and damselflies, birds such as Green Heron and fishes such as Northern Pike.

Functions

Aquatic ecosystems perform many important environmental functions. For example, they recycle nutrients, purify water, attenuate floods, recharge ground water and provide habitats for wildlife. Aquatic ecosystems are also used for human recreation, and are very important to the tourism industry, especially in coastal regions.

The health of an aquatic ecosystem is degraded when the ecosystem's ability to absorb a stress has been exceeded. A stress on an aquatic ecosystem can be a result of physical, chemical or biological alterations of the environment. Physical alterations include changes in water temperature, water flow and light availability. Chemical alterations include changes in the loading rates of biostimulatory nutrients, oxygen consuming materials, and toxins. Biological alterations include over-harvesting of commercial species and the introduction of exotic species. Human populations can impose excessive stresses on aquatic ecosystems. There are many examples of excessive stresses with negative consequences. Consider three. The environmental history of the Great Lakes of North America illustrates this problem, particularly how multiple stresses, such as water pollution; over-harvesting and invasive species can combine. The Norfolk Broadlands in England illustrate similar decline with pollution and invasive species. Lake Pontchartrain along the Gulf of Mexico illustrates the negative effects of different stresses including levee construction, logging of swamps, invasive species and salt-water intrusion

Abiotic Characteristics

An ecosystem is composed of biotic communities that are structured by biological interactions and abiotic environmental factors. Some of the important abiotic environmental factors of aquatic ecosystems include substrate type, water depth, nutrient levels, temperature, salinity, and flow. It is often difficult to determine the relative importance of these factors without rather large experiments.

There may be complicated feedback loops. For example, sediment may determine the presence of aquatic plants, but aquatic plants may also trap sediment, and add to the sediment through peat.

The amount of dissolved oxygen in a water body is frequently the key substance in determining the extent and kinds of organic life in the water body. Fish need dissolved oxygen to survive, although their tolerance to low oxygen varies among species; in extreme cases of low oxygen some fish even resort to air gulping. Plants often have to produce aerenchyma, while the shape and size of leaves may also be altered. Conversely, oxygen is fatal to many kinds of anaerobic bacteria.

Nutrient levels are important in controlling the abundance of many species of algae. The relative abundance of nitrogen and phosphorus can in effect determine which species of algae come to dominate. Algae are a very important source of food for aquatic life, but at the same time, if they become over-abundant, they can cause declines in fish when they decay. Similar over-abundance of algae in coastal environments such as the Gulf of Mexico produces, upon decay, a hypoxic region of water known as a dead zone.

The salinity of the water body is also a determining factor in the kinds of species found in the water body. Organisms in marine ecosystems tolerate salinity, while many freshwater organisms are intolerant of salt. The degree of salinity in an estuary or delta is an important control upon the type of wetland (fresh, intermediate, or brackish), and the associated animal species. Dams built upstream may reduce spring flooding, and reduce sediment accretion, and may therefore lead to saltwater intrusion in coastal wetlands.

Freshwater used for irrigation purposes often absorbs levels of salt that are harmful to freshwater organisms.

Biotic Characteristics

The biotic characteristics are mainly determined by the organisms that occur. For example, wetland plants may produce dense canopies that cover large areas of sediment—or snails or geese may graze the vegetation leaving large mud flats. Aquatic environments have relatively low oxygen levels, forcing adaptation by the organisms found there. For example, many wetland plants must produce aerenchyma to carry oxygen to roots. Other biotic characteristics are more subtle and difficult to measure, such as the relative importance of competition, mutualism or predation. There are a growing number of cases where predation by coastal herbivores including snails, geese and mammals appears to be a dominant biotic factor.

Autotrophic Organisms

Autotrophic organisms are producers that generate organic compounds from inorganic material. Algae use solar energy to generate biomass from carbon dioxide and are possibly the most important autotrophic organisms in aquatic environments. The more shallow the water, the greater the biomass contribution from rooted and floating vascular plants. These two sources combine to produce the extraordinary production of estuaries and wetlands, as this autotrophic biomass is converted into fish, birds, amphibians and other aquatic species.

Chemosynthetic bacteria are found in benthic marine ecosystems. These organisms are able to feed on hydrogen sulfide in water that comes from volcanic vents. Great concentrations of animals that feed on these bacteria are found around volcanic vents. For example, there are giant tube worms (*Riftia pachyptila*) 1.5 m in length and clams (*Calyptogena magnifica*) 30 cm long.

Heterotrophic Organisms

Heterotrophic organisms consume autotrophic organisms and use the organic compounds in their bodies as energy sources and as raw materials to create their own biomass. Euryhaline organisms are salt tolerant and can survive in marine ecosystems, while stenohaline or salt intolerant species can only live in freshwater environments.

Fresh Water

Freshwater refers to the water found in lakes, ponds, streams, and any other body of water other than the sea. It supports a range of plant and animal ecosystems whose composition is shaped by the availability of food, oxygen (O), temperature, and sunlight. Freshwater environments are less extensive than the sea, but they are important centers of biodiversity. This is especially so in dry environments, like deserts, where isolated ponds and streams provide a haven for plants and animals.

Plants and animals living in freshwater would not usually be able to live in saltwater, because their bodies are adapted to a low-salt content. Freshwater ecosystems are vulnerable to water pollution that arises from a range of human activities, from deforestation to urban development. They are also used as water supplies for human use, and sometimes their natural course is diverted for this purpose as, for instance, when a dam is built.

Historical Background and Scientific Foundations

Aquatic, or watery, environments are divided into freshwater and marine. Freshwater has less than one gram per liter of dissolved solids, mainly salts, of which sodium chloride (NaCl) is the most important as far as living organisms are concerned. It is the main source of water for most human uses. Freshwater ecosystems are found in ponds, lakes, reservoirs, rivers, and streams. Estuaries, which are places where a river meets the sea, such as San Francisco Bay, are part freshwater and part marine in their makeup. The diversity of a freshwater ecosystem depends upon temperature, availability of light, nutrients, oxygen, and salinity.

A wide range of plants, animals, and microbes are found in freshwater ecosystems. The smallest are the microscopic plants and animals known as phytoplankton and zooplankton, which form the bottom layer of freshwater food chains. There are also many freshwater invertebrates including worms and insects. Among the freshwater vertebrates, amphibians, such as frogs, live on land and water, while fish are purely water-dwelling. Many species of birds, such as kingfishers and ducks, live on or near freshwater.

Rivers and streams are lotic, or flowing, freshwater environments. Their water flows in one direction and they begin at a source—which could be a spring, lake, or snowmelt—and travel to their mouths, which may be the sea or another river. The water at the source is generally cooler, clearer, and has higher oxygen content than at the mouth. Freshwater fish such as trout are often found near the source. There tends to be more biodiversity in the middle of a river or stream, while the water near the mouth is often murky with sediment that decreases the amount of light and the diversity of the ecosystem. Fish requiring less oxygen, like carp, are found near the mouth of a river. Lotic organisms tend to be small, with flattened bodies, so they are not swept away. Fallen leaves, insects, and other detritus are important food sources.

Rivers and streams carry precipitation to oceans, so there are streams in most localities. There is no sharp boundary between water and land with a stream. There is saturated soil both laterally and vertically beyond the banks of a stream, known as the riparian zone.

Lakes, ponds, and reservoirs are lentic, or layered, systems with generally still water varying in size between a few square feet to thousands of square miles. The surface layer is populated by plankton, protists (single-celled organisms like ameba), and insects. Beneath the surface is the epilimnion, which is relatively warm and sometimes mixed by the wind. The penetration of sunlight through the layer depends upon how much silt is suspended in the water. The layer just above the bottom,

Words to Know

Detritus: Matter produced by decay or disintegration of living material.

Lentic: The vertically layered nature of a lake.

Littoral: The region of a lake near the shore.

Lotic: Flowing water, as in rivers and streams

Wetland: A shallow ecosystem where the land is submerged for at least part of the year.

Known as the hypolimnion, is cold and unmixed. The interface between these two layers, which marks a sudden fall in temperature, is called the thermocline. The bottom layer of a lake, the benthos, is occupied by burrowing worms and snails, with the ecosystem varying depending upon whether the bottom is rocky, muddy, or sandy. Levels of both oxygen and light decline with-depth. Anaerobic microorganisms, which can live without oxygen, often live in the bottom layers of a lake. A blue lake is lower in productivity than a green lake, but when productivity is too high, algal blooms may result with lower oxygen levels. The littoral layer, near the edge of a lake, generally has a large ecosystem with organisms that can use both land and water, such as dragonflies, frogs, ducks, and turtles. Plants such as rushes, reeds, and cattails grow rooted in the bottom sediments of the littoral zone.

Wetlands are important ecosystems that are part aquatic, part terrestrial. They are submerged, either partially or wholly, for at least part of the year. There are various types of wetland, described by their vegetation. Swamps are wetlands with trees, whereas a marsh is a wetland that does not have any trees. A bog contains areas of ground that are saturated with water and its ground is made of a material called peat, which is composed of accumulated and undecayed vegetation. Fens are like bogs, but their water is groundwater, whereas a bog is wet mainly by precipitation. Swamps and marshes are more nutrient-rich and productive than fens and bogs. Wetlands are often rich in biodiversity and are important for breeding and migrating birds and wild flowers. They play an important role by soaking up storm water, preventing flooding by slowing down the rate at which the water reaches river systems. They also act as a filter for agricultural waste, because wetland plants and microbes can detoxify otherwise harmful residues, thereby purifying this water.

Impacts and Issues

Many human activities threaten the health of freshwater ecosystems. For instance, acid rain created from sulfur (S) and nitrogen oxide (NO) emissions turns many lakes and streams acidic, leaving them unable to support various fish species. The building of dams to create hydroelectric power plants blocks the routes of migratory fish such as salmon. Deforestation adds silt to a stream or river and slows it down, which may increase flooding.

Wetlands are among the most vulnerable of ecosystems. Five percent of the land area of the United States is occupied by wetlands that are home to one third of its endangered species. They are often situated near urban areas, where they are an attractive target for draining and development. They often fill with sediment, which is sometimes added to by road building and agricultural runoff, and this can convert them into a terrestrial environment. The United States lost nearly 500,000 acres (200,000 hectares) of wetland each year from the 1950s to the mid-1970s before their ecological importance was realized.

Restoring the water supply is sometimes all that is needed to restore a wetland. One example is the delta where the Tigris and Euphrates rivers empty into the Persian Gulf. This area was home to a group of people called the Marsh Arabs, who resided on floating platforms and lived off the marsh. During the Iran-Iraq war of 1980–1988, Saddam Hussein forced most of these people from their homes and drained the marshes, burning them afterward. After the fall of Hussein, the United Nations and the remaining Marsh Arabs restored the water to the area. Some of the original flora and fauna have started to return, although it will be many years until this wetland is restored to its former state.

References

- Barange M, Field JG, Harris RP, Eileen E, Hofmann EE, Perry RI and Werner F (2010) Marine Ecosystems and Global Change Oxford University Press. ISBN 978-0-19-955802-5

- Introductory-chapter-marine-ecology-biotic-and-abiotic-interactions, marine-ecology-biotic-and-abiotic-interactions: intechopen.com, Retrieved 29 June 2018

- Mann KH and Lazier JRN (2006) Dynamics of marine ecosystems: biological-physical interactions in the oceans Wiley-Blackwell. ISBN 978-1-4051-1118-8

- Freshwater-and-freshwater-ecosystems, energy-government-and-defense-magazines: encyclopedia.com, Retrieved 25 March 2018

- Types-aquatic-ecosystems-6123685: sciencing.com, Retrieved 15 May 2018

Chapter 5

Marine Microbiology

Marine microbiology is the science concerned with the study of microbes that live in saltwater ecosystems, such as the open ocean, estuaries, coastal water and marine surfaces and sediments. The interaction within different communities of microorganisms and their interactions with the environment are also studied under this field. This chapter closely examines the key concepts of marine microbiology, the diverse marine microbes present in such ecosystems such as marine eukaryotes, prokaryotes and viruses, etc.

Marine Microbes

Marine microbes are tiny organisms that live in marine environments and can only be seen under a microscope. They include cellular life forms, such as bacteria, fungi, algae and plankton, along with the viruses that freeload along with them.

More than a billion microorganisms live in each litre of seawater—they dominate the abundance, diversity and metabolic activity of the ocean.

Microbes comprise 98 per cent of the biomass of the world's oceans, supply more than half the world's oxygen and are the major processors of the world's greenhouse gases, which means they have the potential to mitigate the effects of climate change.

Scientists are only just beginning to understand the important environmental roles that microbes play in marine systems—from feeding ecosystems to consuming waste and sequestering carbon. Scientists are investigating several areas where microbial processes are central to issues of immediate concern to the world's coral reefs, including:

- Indicators of ecosystem health and environmental impacts

- Adaptation, acclimatization and evolution of coral reef organisms in the face of global change

- Marine microbiomes and viromes (the ecological communities of microbes and viruses, respectively).

A surprising number of these marine microbes have only recently been discovered, or are known only by their DNA. In some cases, we only know that they are really tiny and have certain types of genes, such as genes necessary for fixing nitrogen. This is one reason that there are no photographs of many of these organisms.

At the broadest level, these microbes can be divided up between three main groups, which are three domains of living things on this planet:

- Bacteria

- Archaea, which look similar to bacteria, but are an entirely separate domain.

- Eukaryotes, a group that includes animals, plants, fungi, protists, and algae.

The microbes listed below are grouped by these three domains. Within each domain, they are grouped according to their roles in nitrogen cycling. We have also included a few microbes that are key players in open-ocean photosynthesis and primary production. These microbes are important for nitrogen cycling in that they "assimilate" nitrogen and convert it into living tissue or "particulate nitrogen."

Several terms are used to describe microbes that perform certain roles in nitrogen cycling, but which may or may not be related to one another. The bacteria and archaea that can convert nitrogen gas into more biologically useful forms such as ammonium are known as "nitrogen fixers" or "diazotrophs." Microbes that convert ammonium to nitrate are called "nitrifying organisms."

Marine Viruses

Viruses in the Sea

Viruses can be found wherever life is found and in seawater they outnumber Bacteria, Archaea, and Eukarya combined by an order of magnitude or more (Wommack and Colwell, 2000;Suttle, 2007). The concentration of virions in surface seawater in the open ocean is on the order of 5–10 billion per liter. Their concentration declines with depth, but viruses are present in the deepest waters sampled (>4000 m). The concentration of virions in surface marine sediments is typically even higher than in the immediate overlying waters. As in the water column, viruses become less abundant with depth in the sediment, but remain detectable to depths >100 m below the seafloor and in sediments with estimated ages of hundreds of thousands of years or more.

Virions suspended in seawater are a component of the marine plankton and are therefore often referred to as virioplankton. Those in sediments are sometimes referred to as viriobenthos. Plankton ecologists find it convenient at times to refer to plankton according to which of seven different size categories they belong. Because of their small size, most of the viruses in seawater are members of the smallest named size category, the femtoplankton (0.02–0.2 μm), but larger viruses are also part of the picoplankton (0.2–2 mm), a size class that also includes almost all of the bacteria and archaea and the smallest protists.

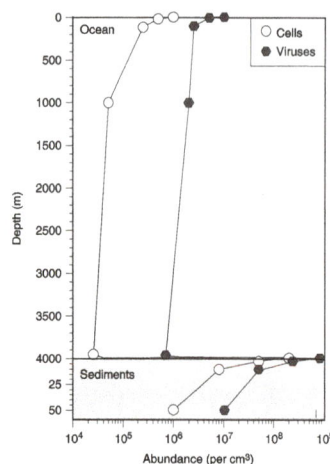

Idealized plot of the concentrations of viruses and cells in open ocean seawater and sediments.

Sediments are indicated by the shaded area. Sediment depth (in bold) starts over from zero at the sediment–water interface and is plotted at a different scale for clarity. "Cells" refers to all cells, but these numbers are dominated by bacteria and archaea. The numbers shown are approximate and concentrations at a given depth can vary by an order of magnitude depending on location. Concentrations tend to be higher in areas of high productivity such as the coastal zone.

Viral infections have been reported for a broad range of marine organisms and these infections have wide-ranging effects on everything from evolution, ecology, and ocean biogeochemistry to human health and economics. Viral infections of phytoplankton, the foundation of the marine food web, affect how much of the ocean's primary production is transferred to higher trophic levels. Large increases in phytoplankton concentrations (blooms) are a common phenomenon in the ocean and these periodic in-creases in primary production serve as a critical food supply for life in the sea. Viruses also spread more easily through populations at high density, however, and phytoplankton blooms can be decimated by the rapid spread of viral infection. Lysis of phytoplankton decreases the efficiency with which they are consumed by grazers and much of the primary production is instead decomposed by bacteria. Bacteria are themselves susceptible to viral infections and viral lysis of bacteria likewise diverts material and energy from protistan grazers, which favors further growth and respiration by other uninfected bacteria. It thus appears that the net effect of lytic viral activity in the marine food web is to decrease the amount of carbon that reaches higher trophic levels (e.g., fish) while increasing the amount of carbon, which is respired by the bacteria.

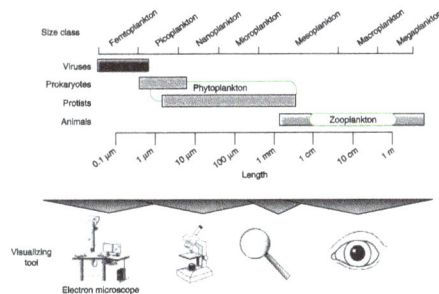

Viruses as members of the marine plankton.

As a group, viruses are the smallest and most abundant members of the plankton and an electron microscope is required to reveal their details. Plankton are often named by the size category to which they belong, but may also be referred to in functional categories such as virioplankton, phytoplankton, and zooplankton.

From the perspective of humans, viruses could be lamented as reducing the efficiency of wild fisheries, but by how much is unknown. In any case, there is little to be done about the persistent, ubiquitous influence of viruses on the ocean's food web; it is an inescapable feature of the natural cycle of life in the sea. However, viruses are a legitimate human concern with tangible economic consequences when they cause death and disease among farmed marine species such as shrimp and fish. As wild fisheries become increasingly threatened by overexploitation, humans are relying more heavily on aqua- and mariculture as a source of protein. Intensive mariculture involves growth of single species at high densities, which, unfortunately, is a condition under which viral and other diseases can easily spread within the farm and from the farm to wild populations. The economic losses resulting from viral diseases in aquaculture are estimated to be in the billions of dollars annually so there is a strong incentive to find solutions to combat viral diseases in commercially exploited marine species.

Life in the sea is very diverse, with representation of most of the major types of life on the planet. As one might then expect, the diversity of viruses in the sea is also broad. Among viruses infecting the marine organisms, members of all Baltimore groups except for Group VII have been found so far.

Taxonomic Viruses

Specific genera within a family are noted in some cases where it seems of particular interest. The taxonomic groups represented are, for the most part, restricted to those explicitly mentioned in the most recent report of the ICTV, but some information is included from other sources and proposals that have been accepted since the report was completed. For the purposes of this presentation, a virus was considered marine if it infects an organism commonly referred to as marine. This includes organisms for which the entire life cycle is completed in the ocean, but also sea turtles (marine reptile), a variety of birds such as gulls, terns, cormorants, and pelicans (marine birds) as well as seals and sea lions (marine mammals). The general characteristics of the virions in each taxon and the types of marine organisms they infect are noted.

Group	Genome		Taxonomy			Marine hosts	
	Type	Segments	Order	Family	Subfamily	Types	Names
I	dsDNA	Mono	Caudovirales	Myoviridae	–	B	Cyanobacteria and proteobacteria
				Podoviridae	–	B	Cyanobacteria and proteobacteria
				Siphoviridae	–	B	Cyanobacteria and proteobacteria
			Herpesvirales	Alloherpesviridae	–	V	Fish
				Herpesviridae	Alphaherpesvirinae	V	Turtle
					Gammaherpesvirinae	V	Seal
				Malacoherpesviridae	–	I	Oyster
			Unassigned	Adenoviridae	–	V	Fish
				Corticoviridae	–	B	Bacteria
				Iridoviridae	–	V	Fish
				Mimiviridae	–	Pr	Heterotrophic protist
				Nimaviridae	–	I	Shrimp and crab
				Papillomaviridae	–	V	Porpoise and dolphin
				Phycodnaviridae	–	Pr	Micro- and macroalga
				Poxviridae	–	V	Marine mammals and birds
				Unassigned	–	Pr	Microalga

II	ssDNA	Mono	Unassigned	Anelloviridae	–	V	Sea lion
				Circoviridae	–	V	Herring gull
				Inoviridae	–	B	Proteobacteria
				Microviridae	Gokushovirinae	B	Proteobacteria
				Parvoviridae	–	I	Shrimp
				Unassigned	–	Pr	Microalga
III	dsRNA	Bi	Unassigned	Birnaviridae	–	I, V	Molluscs, crustaceans, and fish
		Multi	Unassigned	Reoviridae	Spinareovirinae	I, V	Bivalves and fish
					Sedoreovirinae	Pr, I	Microalga and crab
IV	ssRNA (+)	Mono	Nidovirales	Roniviridae	–	I	Shrimp
		Mono	Picornavirales	Dicistroviridae	–	I	Shrimp
				Marnaviridae	–	Pr	Microalga
				Unassigned	–	Pr	Labrynthulid and microalga
		Mono	Unassigned	Astroviridae	–	V	Sea lion and dolphin
				Caliciviridae	–	V	Seals and whales
		Bi		Nodaviridae	–	V	Fish
		Mono		Togaviridae	–	V	Fish and mammals
V	ssRNA (–)	Mono	Mononegavirales	Paramyxoviridae Rhabdoviridae	Paramyxovirinae –	V V	Seals and whales Fish
		Multi	Unassigned	Orthomyxoviridae	–	V	Fish, sea birds, seals, and whales
VI	ssRNA (+)	Mono	Unassigned	Metaviridae	–	I, V	Sea urchin and fish

Group I: dsDNA Viruses

The dsDNA virus isolates known to infect marine organisms are assigned to 13 of the 30 families in Baltimore Group I. Three families are members of the order Caudovirales, another three are members of the order Herpesvirales, and six families are not assigned to an order. A number of viruses infecting marine plankton have also not yet been assigned to a family.

Order Caudovirales

The viral order Caudovirales contains three related families of tailed viruses, all of which infect bacteria or archaea. Members of this order have nonenveloped capsids with icosahedral symmetry, but these may be isometric or prolate with diameters ranging from around 40–145 nm in diameter. Helical tails (10–355 nm long and 5–18 nm wide) aid in attachment to the host cell and serve as a conduit through which the viral genome is delivered to the interior of the cell. The tail structure serves as a taxonomic character that divides the order into the families Myoviridae, Podoviridae, and Siphoviridae can be found in all three of these families, but marine isolates reported to date are restricted to those infecting bacteria, probably because isolates of marine archaea remain so rare. Members of all three families are very common in seawater.

Family Myoviridae

Viruses in this family are characterized by having a long (50 to more than 200 nm) hollow tail with a neck and a contractile sheath. Because of the sheath, the tails of these viruses are thicker (16–18 nm) than those of other caudovirads. Myovirids that infect marine bacteria, particularly bacteria in the phylum Gammaproteobacteria, have been classified, but many more marine isolates are known. These include a large number of isolates infecting cyanobacteria in the genera Prochlorococcus, and Synechococcus (Sullivan et al., 2010). Although there is considerable overlap, the genomes and capsids of myovirids tend to be larger than those of the other two families in this order.

Family Podoviridae

Viruses in this family are characterized by the presence of a short, noncontractile tail that is 10–48 nm long and about 8 nm wide. Podovirids appear to be the most numerous of all tailed phages in the marine environment. Podovirids infecting marine bacteria in the phyla Alpha- and Gammaprotoebacteria have been officially classified by the ICTV, but many more isolates infecting a range of bacteria exist, including podovirids infecting Prochlorococcus and Synecho- coccus.

Family Siphoviridae

Viruses in this family have a long, noncontractile tail that is around 100 to > 4300 nm long, and 5–10 nm wide. These long, unsheathed tails often appear to be quite flexible, displaying bends and curves in electron micrographs.

Order Herpesvirales

Members of the order Herpesvirales have a large icosahedral capsid enclosed in a tegument and a lipid bilayer membrane. The genome ranges from 125,000 to 295,000 bp. Capsids are around 125 nm, but the entire virion is closer to 200 nm in diameter. These viruses are very wide- spread in the vertebrates. Examples of herpesvirads infecting marine vertebrates are known in all three families of the order.

Family Alloherpesviridae

Members of this family include those that infect fish and amphibians. Marine and estuarine representatives are in the genus Ictaluvirus and include those that infect catfish, sturgeon, salmon, and cod.

Family Herpesviridae

This family consists of genetically related herpesvirids that infect reptiles, birds, and mammals. Marine representatives have been assigned to two of the three subfamilies (Alpha- and Gammaherepesvirinae), but are not yet assigned to a genus. Marine hosts include sea turtles, seals, sea lions, dolphins, and whales.

Family Malacoherpesviridae

This new family consists of a single genus (Ostreavirus) and species (Ostreid herpesvirus 1). This virus infects oysters and is the only herpesvirad identified so far that infects an invertebrate.

Families not Assigned to an Order

Family Adenoviridae

Members of this family are nonenveloped and have a capsid 70–90 nm in diameter with icosahedral symmetry and a genome between 25,000 and 50,000 bp. Adenovirids infect a wide variety of vertebrates. There are a number of reports of infections by adenovirids in marine mammals (pinnipeds and cetaceans) and in gulls, but none have been isolated and officially classified. Sea lions may be a host for canine adenovirus. The only reported fish adenovirus is in the genus Ichtadenovirus with the sole representative being Sturgeon ade- novirus A, which infects white sturgeon. Although not truly marine, the sturgeon habitat includes brackish coastal waters.

Family Corticoviridae

This family is represented by a single genus (Corticovirus) and a single species Pseudoalteromonas phage PM2) that infects a marine bacterium. The virion has an icosahedral capsid 57 nm in diameter, no tail, and a lipid bilayer membrane interior to the capsid. The genome in the virion is supercoiled with a length of 10,079 bp. Although only one species has been classified at present, related sequences have been detected within the genomes of a variety of bacteria suggesting that they could be prevalent in the marine environment.

Family Iridoviridae

Members of this family have a large icosahedral capsid (120–200 nm diameter) and a large genome (around 100,000 to 200,000 bp). They infect invertebrates (arthropods) and vertebrates (amphibians, reptiles, and fish). Those infecting invertebrates are nonenveloped while those infecting vertebrates acquire a lipid envelope that derives from the host cell membrane during virion budding. Members of two of the five genera in this family (Lymphocystivirus and Megalocytivirus) infect marine fish.

Family Mimiviridae

This is a relatively new family of viruses, the members of which have unusually large genomes and icosahedral capsids. The first representative to be fully sequenced, Acanthamoeba polyphaga mimivirus (APMV) infects a freshwater amoeba. At the time of its discovery, it was the largest known virus with a genome of 1,181,549 bp and a capsid diameter of 500 nm. The virus is densely covered with hair-like fibrils that extend its effective diameter to 750 nm. Recently, another virus related to APMV with a larger genome (1,259,157 bp), but smaller capsid (440 nm) was isolated from a marine sample using several freshwater Acanthamoeba species as hosts. The natural host of this new isolate, named Megavirus chilensis, is unknown, but is presumed to be a marine or estuarine protist. This is the largest viral genome known at the present time.

Family Nimaviridae

This family consists at present of a single genus (Whispovirus) and species (White spot syndrome virus). Virions consist of a rod- shaped nucleocapsid (approximately 65–70 x 300–350 nm) enclosed in a tegument and a trilaminar envelope. The overall virion dimensions are 120–150 nm in diameter and 270– 290 nm in length. The virion is unusual in having a thread-like appendage that gives it the superficial appearance of a bacterium with a polar flagellum.

Family Phycodnaviridae

The members of this family that have been officially classified so far have large icosahedral capsids (around 100–200 nm diameter) with no outer envelope and genome sizes ranging from approximately 100,000 to 500,000 bp. These viruses infect photosynthetic protists (micro and macroalgae) in both freshwater (genus Chlorovirus) and the marine environment (genera Coccolithovirus, Phaeovirus, Prasinoviruses, Prymnesio- viruses, and Raphidoviruses). Members of the genus Phaeovirus infect filamentous brown algae, but all others infect unicellular plankton.

Family Poxviridae

The capsids of poxvirids are enveloped and brick shaped or ovoid with tubular surface processes. They vary from 200 to 450 nm in length, 140–280 nm in width, and 140–280 nm in thickness. The genomes are linear dsDNA ranging from 130,000 to 380,000 bp. Members of this family infect a wide range of hosts. The poxvirids infecting vertebrates are in the subfamily Chordopoxvirinae, which contains nine genera. The subfamily Entomopoxivirinae comprises three genera, the members of which infect various insects. The family is well known for the serious human disease caused by one of its notorious members, small pox (now eradicated). In the marine realm, poxviruses have been identified in penguins, seals, sea lions, dolphins, and whales, but none of these have been officially classified.

Unassigned

Several new viruses infecting marine heterotrophic and phototrophic protists have been isolated and described, for which the taxonomic positions remain unclear. A large (300 nm) virus, CroV, that infects Cafeteria roenbergensis is the first isolate that infects a phagotrophic marine protist. Based on the large genome size (750,000 bp) and specific sequence of the DNA polymerase of CroV, it may belong to the family Mimiviridae. A number of other virus isolates infecting photosynthetic marine protists (Chrysochromulina ericina virus 01B, Phaeocystis pouchetii virus 01, and Pyramimonas orientalis virus 01B) have capsids closer in size to phycodnavirids (140–220 nm), but their DNA polymerase genes place them closer to APMV and CroV, suggesting they may also be members of the Mimiviridae.

A large virus (200 nm diameter, 350,000 bp genome) was isolated that infects the bivalve-killing dinoflagellate, Heterocapsa circularisquama . The DNA poly- merase sequence of this isolate is very similar to that of African swine fever viruses in the genus Asfivirus, which is the sole genus in the family Asfarviridae. It is, therefore, highly di- vergent from other dsDNA viruses infecting phytoplankton and a new genus name, Dinodnavirus, has been proposed for this virus.

Group II: ssDNA Viruses

Family Anelloviridae

Members of this family have a nonenveloped icosahedral capsid approximately 30 nm in size and a circular genome ranging in size from 2000 to 3900 nt. They appear to be widespread in mammals, but their involvement in specific diseases remains unproven. Novel anellovirids have recently been detected by metagenomic analyses of lung tissue from seals and sea lions, but these have yet to be approved as species.

Family Circoviridae

This family is comprised of two genera (Circovirus and Gyro virus) known to infect birds and swine, although distantly related viruses have been described in primates. The virions are nonenveloped and appear spherical with icosahedral symmetry. The capsid diameter is 20.5–25 nm and the gen- ome ranges from 1775 to 2300 nt. The only classified member so far that might be considered marine is the gull circovirus in the genus Circovirus.

Family Inoviridae

Virions in this family are filaments or rods with helical symmetry with a circular genome ranging from 4500 to 12,400 nt in length. Those in the genus Inovirus are long (700–900 nm) and thin (7 nm) and infect gammaproteobacteria, while those in the genus Plectrovirus (10–15 nm wide and 70–280 nm long) infect Gram-positive bacteria. Among the known hosts of inovirids are members of the bacterial genus Vibrio, which are common in coastal estuarine and marine waters. Inovirids cause persistent infections and are responsible for lysogenic conversion. The inovirid, Vibrio cholera CTXø, for example, codes for an enterotoxin that converts Vibrio cholerae into a human pathogen.

Family Microviridae

Virions in this family have a small, nonenveloped capsid ap- proximately 25–27 nm in diameter and a circular, positive- sense genome ranging in size from 5300 to 6100 nt (genus Microvirus) or from 4400 to 4900 nt (genera Chlamydiamicrovirus, Bdellomicrovirus, and Spiromicrovirus in the subfamily Gokushovirinae). The hosts for members of the genus Bdellomicrovirus, are bacteria in the genus Bdellovibrio. Bdellovibrio spp. are predatory bacteria found in a range of freshwater, estuarine, and marine habitats. All strains of bdellomicroviruses isolated so far have been from freshwater. However, there is evidence from metagenomics that additional members of the Gokushovirinae are widespread in the ocean. Sequence comparisons indicate that these viruses are distantly related to known chlamydiamicroviruses and bdellomicroviruses, but their hosts have not yet been identified.

Unassigned

Several genetically similar ssDNA viruses infecting diatoms in the genus Chaetoceros have been isolated. The virions are nonenveloped and have an icosahedral capsid ranging in size from 30 to 38 nm and a genome around 6000 nt in length. Although these viruses do not yet belong to an official family, a new genus, Bacilladnavirus, has been approved to accommodate this group of viruses.

Group III: dsRNA Viruses

Family Birnaviridae

There are four genera in the family Birnaviridae. The icosahedral particles of birnavirids are 65 nm in diameter, none- nveloped and are composed of a single capsid protein. The genome consists of two linear segments of dsRNA approximately 5900–6900 bp in total length that encodes a poly- merase and a large structural polyprotein that is cleaved post-translationally. Birnavirids infect rotifers, insects, fish, and birds and are globally distributed. In the marine en- vironment, aquabirnaviruses cause disease in salmonids, amberjack (yellowtail), bream, and cod among other fishes, and even molluscs and crustaceans.

Family Reoviridae

The family Reoviridae has fifteen genera divided among two subfamilies, the Spinareovirinae that include viruses that have turret-like proteins projecting from their capsid, and the Sedoreovirinae, the reovirids without turrets. The virions in this family are nonenveloped icosahedrons that range from 49 to 100 nm in diameter. Reovirids are distinctive in that their capsid can be composed of as many as three concentric protein shells that surround from 9 to 12 segments of dsRNA, that together range between 18,500 and 33,300 bp in total length. The viruses in this family infect a wide range of multicellular organisms including protists, fungi, plants, insects, reptiles, and humans. The dsRNA virus Micromonas pusilla reovirus (MpRV) from the genus Mimireovirus was one of the first marine RNA viruses isolated and it infects a marine protist . The host, M. pusilla, is a small (<2 mm), autotrophic flagellate considered the most abundant picoeukaryote in the global oceans. Species of bivalves and finfish are subject to infection from reovirids in the genus Aquareovirus, while orbiviruses have been found in seabirds (e.g., murres, puffins, and razorbills) and cardoreoviruses in crabs.

Group IV: Positive-Sense ssRNA Viruses

Order Nidovirales

The order Nidovirales is divided into three families, two sub- families and seven genera. Virions in this order range widely in size from 50 to 200 nm, but all are enveloped. The genome of a nidovi- rad is a single, linear molecule of RNA that ranges in size from 12,700 to 29,100 bp and has a 50 cap and a poly(A) tail that extends from the 30 end. The gene order is highly conserved, with open reading frames (ORFs) nearest the 50 end encoding the replicase, followed by 3–12 structural proteins. The genomes of viruses in the Nidovirales contain a characteristic - 1 frameshift that is involved in the regulation of viral protein production during replication. Members of this order are the largest RNA viruses known to date, both in genome and particle size.

Family Roniviridae

The genus Okavirus is the only genus in the family Roniviridae. The virions of ronivirids are enveloped, bacilliform in shape, vary in size from 150 to 200 nm, and are studded with spike-like glycoprotein projections that have bulbous distal ends. The genome of these viruses has features typical of the order Nidovirales and can be quite large for an RNA virus, exceeding

26,000 nt in length. The genome configuration, particularly in the structural genes, as well as the absence of a discrete structural membrane protein gene, distinguishes ronivirads from other viruses in the order. Viruses in the family Roniviridae infect the giant tiger shrimp and the blue shrimp.

Order Picornavirales

The sprawling order of the Picornavirales contains five families, two subfamilies, 24 genera, and approximately 120 species of viruses. Picornavirads are "picorna-like" in that they have similar features to viruses in the family Picornaviridae. Members of the order have a nonenveloped icosahedral capsid that ranges in diameter from 22 to 30 nm. The linear, ssRNA genome is 7000–15,500 nt in length and encodes a polyprotein that is post-translationally processed. Picornavirads also have in common a conserved nonstructural gene order and a small protein covalently linked to the 50 end and a 30 end poly(A) tail. These viruses infect a wide range of eukaryotes, including humans. Picornavirads of note include human enterovirus C, which causes poliomyelitis, human rhinovirus, responsible for the common cold, and Israeli acute paralysis virus, which has been linked to honey bee colony collapse disorder.

Family Dicistroviridae

The family Dicistroviridae has two recognized genera, the cri- paviruses and the aparaviruses. The particles of dicistrovirids are nonenveloped icosahedra approximately 30 nm in size. The genome is a single molecule of linear RNA that can be from 8500 to 10,000 nt in length and encodes two poly- proteins, or cistrons (ergo the name of the family). Dicistrovirids infect only invertebrates, including aphids, bees, and butterflies. In the marine environment, Taura syndrome virus (TSV), an aparavirus, infects penaeid shrimp. TSV epidemics have been responsible for high rates of mortality in shrimp farms.

Family Marnaviridae

Heterosigma akashiwo RNA virus (HaRNAV) is the only member of the Marnaviridae. Heterosigma akashiwo, the host of HaRNAV, is a small, motile photosynthetic protist that is a common member of the temperate coastal phytoplankton community. During times when surface waters are stratified as a result of higher sea surface temperatures and low winds, Heterosigma akashiwo is capable of forming immense blooms that can cause extensive fish kills. These toxic blooms affect the fish farming industry in the fjords of western Canada, and Scandanavia in particular. In 2003, Tai and colleagues isolated HaRNAV, the first RNA virus reported to infect a protist . HaRNAV has a nonenveloped, icosahedral capsid that is approximately 25 nm in diameter. The burst size of HaRNAV is estimated to be 20,000 viruses per cell. The linear ssRNA genome of HaRNAV is 8600 nt in length and is monocistronic – unlike AuRNAV, the bacillar- naviruses, TSV, and HcRNAV. However, the HaRNAV genome has the same gene order as the other marine ssRNA viruses with the nonstructural genes nearest the 50 end of the genome followed by the structural genes. Phylogenies based on the conserved domain sequences from HaRNAV and the corresponding regions of other viruses place HaRNAV outside previously characterized virus taxa within the order Picornavirales.

Proposed Genus Bacillarnavirus

The genus Bacillarnavirus has not been assigned to a family within the order Picornavirales. It has three recognized mem- bers thus far, Rhizosolenia setigera RNA virus (RsetRNAV), Chaetoceros tenuissimus Meunier RNA virus (CtenRNAV), and Chaetoceros socialis f. radians RNA virus (CsfrRNAV) all of which infect marine diatoms. Diatoms are widely distributed, highly diverse, and are among the most abundant eukaryotes in the ocean. They form a large fraction of the total marine primary producers and are thus significant contributors to the cycling of energy in the ocean. The hosts of CtenRNAV and CsfrRNAV are members of the genus Chaetoceros, a group of chain-forming diatoms that can form blooms associated with fish kills. The two viruses have icosahedral symmetry and have ssRNA genomes that are dicistronic and approximately 9500 nt in length. CtenRNAV has an estimated capsid diameter of 31 nm, while that of CsfrRNAV is only 22 nm. A striking difference between the two viruses is that CtenRNAV has an estimated burst size of 10,000 virions while CsfrRNAV produces approximately 66 per lytic event. RsetRNAV was the first virus brought into culture that infects a diatom. Like CtenRNAV and CsfrRNAV, RsetRNAV has an icosahedral capsid (approximately 32 nm in diameter), replicates in the cyto- plasm and has a positive-sense ssRNA dicistronic genome ap- proximately 8900 long. The genomes of all three RNA viruses have a similar gene order, with the structural genes nearest the 50 end, followed by the helicase, protease, and replicase and terminating in a poly(A) tail. Alignments of the replicase genes of these viruses clearly demonstrate that they are closely related and form a well- supported monophyletic clade.

Proposed Family Labyrnaviridae

Aurantiochytrium RNA virus (AuRNAV) is the only member of the family Labyrnaviridae. AuR-NAV infects the thraustochytrid Aurantiochytrium sp. Thraustochytrids are single-celled marine organisms that were once thought to be fungi, but are now recognized as a class of protists. These organism are capable of feeding directly through absorption and are involved in the degradation of detritus. Like HcRNAV and the bacillarnaviruses, AuRNAV has a small (roughly 25 nm), nonenveloped icosa- hedral capsid. AuRNAV replicates in the cytoplasm and has a phenomenal burst size of approximately 60,000 particles per lytic event. The genome of AuRNAV is a single molecule of RNA 9000 nt in length with a poly(A) tail. The three ORFs of the AuRNAV genome encode the capsid proteins, nonstructural genes and a protein of unknown function. Interestingly, AuRNAV produces subgenomic RNA during replication. Although the specific mechanism is unknown, this suggests that the AuRNAV genome does not possess an internal ribosome entry site structure that is characteristic of other picornavirads.

Families Not Assigned to an Order

Proposed Family Alvernaviridae

The family Alvernaviridae has a single representative at present, Heterocapsa circularisquama RNA virus (HcRNAV), which infects the eponymous dinoflagellate. The dinoflagellates are a diverse group of unicellular flagellates that are capable of de- riving energy in a number of different ways, including through parasitism, photosynthesis, and predation. Dinoflagellates are found in marine and freshwater, sediments, snow and ice, and even inside other organisms. They are an important food source for plankton, a critical component of coral reefs, and can periodically form potent harmful algal blooms. Heterocapsa circularisquama is an armored, free-living protist that can form extensive blooms and these blooms have resulted in severe financial losses to the aqua- culture industry. HcRNAV particles do not have an envelope and are approximately 30 nm in diameter. The linear RNA genome of this virus is only 4400 nt long. The two ORFs of the HcRNAV genome encode a protease, a replicase, and a single capsid protein. During the final stages of HcRNAV replication, the virus forms tightly packed crystalline arrays within the cell that result in the release of an estimated 20,000 viral particles during lysis.

Family Astroviridae

The family Astroviridae includes two genera. The capsids of astrovirids have nonenveloped, spherical particles that range in size from 28 to 30 nm and are covered in mushroom-like projections involved in virus attachment and entry. The genome of these viruses is a single linear molecule of RNA that is generally from 6400 to 7700 nt long and terminates in a poly(A) tail. Like the calicivirids, the astrovirids produce a subgenomic RNA that encodes the structural proteins. Astrovirid infections cause gastroenteritis in a variety of birds and mammals, including humans. In the marine environment, members of the Mamastrovirus genus infect Bottlenose dolphins, and California and Steller sea lions.

Family Caliciviridae

There are five genera of viruses within the family Caliciviridae. Viruses in this family produce nonenveloped, icosahedral virions 27–40 nm in diameter that has distinctive cup- shaped depressions. The genome of these viruses is linear, ranges from 7300 to 8300 nt long and has a poly(A) tail. Unlike some other taxa of group IV viruses, the calicivirids make a subgenomic RNA that encodes the structural genes during replication. Humans, pigs, rodents, and cats, among other vertebrates are all infected by calcivirids. Viruses in the genus Vesivirus have been found to infect diverse marine mammals, including sea lions, cetaceans, and walruses. Vesiviruses exhibit a broad cell tropism and a broad host range. For example, vesiviruses isolated from marine mammals are capable of infecting pigs and humans.

Family Nodaviridae

There are two genera in the family Nodaviridae. Viruses in this taxa are generally spherical in shape, (from 25 to 33 nm in diameter) and lack an envelope and discernible surface structure. The genome of nodavirids consists of two molecules of RNA (approximately 4500 nt in total length) that have 50 caps and lack poly(A) tails. Nodavirids are capable of infecting a variety of insect species including two species of Aedes mosquitos, in which they cause paralysis and death. Viruses in the genus Betanodavirus infect and cause disease in marine fish. Examples include flounder, grouper, and pufferfish, among others. In addition, several nodavirid-like viruses not yet approved by the ICTV infect prawns.

Family Togaviridae

This family comprises two genera, the alphaviruses and rubiviruses. While the Alphavirus genus has approximately 40 recognized members, the Rubella virus is the only member of the Rubivirus genus. The particles of these viruses are enveloped, spherical in shape and have an approximate diameter of 70 nm. The genome of a togavirid is a single strand of RNA that is from 7700 to 11,800 nt in length and has a 50 cap and a poly(A) tail. While most togavirids are transmitted via an arthropod intermediary, transmission of the alphaviruses that infect marine mammals and fishes do not appear to require a vector. Alphaviruses have been isolated that infect elephant seals, Atlantic salmon, and rainbow trout.

Group V: Negative-Sense ssRNA Viruses

Order Mononegavirales

The Mononegavirales is a large order of negative-sense single- stranded viruses that includes four families, two subfamilies, and 16 genera. Although the morphology of these viruses can vary widely (ranging from 80 to 805 nm in size), all virions are enveloped and most possess spike-like structures that are involved in attachment to and fusion with the host cell. The genomes of mononegavirads are linear and nonsegmented, range from 8900 to 19,100 kb in size, lack a 50 cap and a poly(A) tail, have a characteristic gene order (wherein from the 30 end, core protein genes are followed by envelope genes and then the polymerase), and encode an RNA-dependent RNA polymerase whose sequence is highly conserved. During replication, each cistron is transcribed utilizing a stop-start mechanism starting at a single promotor region located on the 30 ends. The Mononegavirales is a taxon of particular relevance to human health, because it includes virulent human pathogens such as the Ebola virus, rabies virus, and measles and mumps viruses, but also includes many representatives infecting marine organisms.

Family Paramyxoviridae

The paramyxovirids are viruses of vertebrates that are re- sponsible for notorious diseases in humans and animals such as measles, mumps, pneumonia, rinderpest, and Newcastle disease. The family is organized into two subfamilies comprised of seven genera. The virions of paramyxovirids are spherical in shape and approximately 150 nm in diameter. The genome of these viruses is a single molecule of negative- sense RNA that can be from 13,300 to 18,200 kb in length. All genomes of paramyxovirids encode a protein, creatively named "F," which causes the fusion of the virus and cell membranes at neutral pH. An interesting characteristic of the replication of paramyxovirids is the "rule of six." This rule alludes to the fact that the efficiency of RNA replication in these viruses is severely diminished if the number of nucleotides in the RNA template being replicated is not exactly a multiple of six. It is believed that the precise packaging requirements of the virus drive this surprising requirement. The Cetacean morbillivirus and the Phocine distemper virus are species in the genus Morbillivirus that infect marine mammals, while the Newcastle disease virus has been isolated from the common murre.

Family Rhabdoviridae

More than 200 different strains of rhabdovirids have been isolated from plants, invertebrates and vertebrates. The most distinctive feature of this family of viruses is that they all possess a bullet or rod-like morphology that can be from 100 to 430 nm in size. The genome of a rhabdovirid is a single molecule of negative-sense RNA that can be from 11,000 to 14,900 kb in length. Viruses in the genus Novirhabdovirus infect diverse fishes, including salmon, flounder, and eels. Two novirhabdoviruses of note, infectious hematopoietic necrosis virus and viral hemorrhagic septice-mia virus, are responsible for the mortality of significant numbers of wild and farm-raised salmon each year. The rhabdovirid Dolphin rhabdovirus-like virus was isolated from a white-beaked whale in the North Sea.

Families Not Assigned to an Order

Family Orthomyxoviridae

The virions of orthomyxovirids are enveloped, spherical in shape, and range in diameter from 80 to 120 nm. They contain a negative-sense RNA genome composed of six to eight segments of different length that are in total from 10,000 to 14,700 kb in size. Because of their ubiquity and virulence, orthomyxovirids have been intensely studied since the early twentieth century. This research has resulted in an unprecedented understanding of the mechanisms of zoonosis and evolutionary history of these viruses. Analyses of genomic data suggest that wild aquatic birds may be the primary reservoir of the orthomyxovirids that now infect a wide range of vertebrates, including the human-infecting H1N1 strain responsible for the worldwide human pandemic of 1918. Of the five genera in the family, the genera Isavirus and Influenzavirus A have viruses that infect fishes and marine mammals.

Family Bunyaviridae

The family Bunyaviridae comprises five genera whose viruses infect vertebrates and plants primarily via an arthropod vector. Among other distinguishing features, bunyavirids have a genome comprised of three unique molecules of negative or ambisense ssRNA, that range in total size from 11,400 to 19,200 kb, that encodes four structural proteins. The particles of these viruses are pleomorphic and range in size from 80 to 120 nm. The terminal sequences of the three-genome segments are highly conserved among bunyavirids and play an important role during viral replication. Most bunyavirids are capable of replicating in both arthropods, where little cytopathogenicity occurs, and vertebrates, where they are cytolytic. Mourilyan virus is a bunya-like virus that infects two penaied prawns. Additional bunyavirids have been isolated that infect seabirds, including murres and puffins.

Group VI: Positive-Sense ssRNA Viruses that Replicate through a DNA Intermediate

Family Metaviridae

Metavirids blur the line between virus and retrotransposon in that only one of the species of virus in this family produces infectious extracellular particles (these virions are irregular in shape and range in size from 50 to 100 nm). Oddly, encapsidated particles are produced within the cytoplasm of the host cell and in most cases do not exit. Nevertheless, viruses in this family encode the hallmark genes (major capsid protein, reverse transcriptase, and integrase), share the same gene order, and have a similar genome length to retroviruses (5200–7600 kb). There are two known marine metavirids that infect the same species of pufferfish, the Takifugu rubripes Sushi virus (TruSusV) and Fugu rubripes Suzu virus (FruSuzV), classified in the genera Metavirus and Semotivirus, respectively.

Phenotypic and Genotypic Diversity of Marine Virioplankton

Detailed characterizations of cultivated viruses have been essential for developing a coherent viral taxonomy. This structure has brought order to the bewildering complexity of the viral world, but viral diversity as viewed through the lens of taxonomy has its limitations. The stringent requirements for establishing official taxonomic classifications mean that the process is slow and deliberate. The rate at which novel viruses can be identified far outstrips the rate at which they can be properly classified. Taxonomy also does not concern itself with relative abundance or biogeography, which is an issue of interest to ecologists. Microbial ecologists investigating viruses in the ocean have therefore been somewhat less concerned with taxonomic classifications than with practical measures of diversity. Some studies rely on cultivation (e.g., establishing host ranges), but many others have focused on quantifying abundance and diversity (phenotypic and genotypic) using cultivation-independent methods.

Host Range

The low-throughput of cultivation is a bottleneck in the characterization of viral communities, but the ability to propagate a virus in the lab allows one to collect ecologically relevant information that cannot be obtained any other way. One can determine, for example, in how many different types of cells the virus can replicate. Investigations of viruses with different host ranges can provide insight into the fitness costs that must necessarily be incurred to offset the benefits of being able to infect a broader range of cell types. Although viruses are often classified as having a broad or narrow host range, such a semiquantitative conclusion is necessarily conditional, since not every possible host can be screened. Despite that limitation, host range analysis has been very enlightening, because it can be exquisitely sensitive in discriminating between strains and it does not require sophisticated or expensive technology. To perform this analysis, each viral isolate is tested for its ability to lyse a suite of organisms. Since viruses tend to be more or less host-specific, the suite of organisms tested should primarily consist of different strains of the same species or different species in the same genus, but testing more distantly related organism allows one to identify viruses with particularly broad host range. The more organisms one tests, the greater one's chance of discriminating between closely related viruses. Use of this technique with marine bacteria has shown that the number of viruses infecting any given genus of marine bacterium is extraordinarily large.

Morphology

The subnanometer resolving power of the transmission electron microscope (TEM) allows one to discriminate fine structural details of individual virions. This power has been exploited to conduct qualitative surveys of marine viral diversity that hint at an extraordinary richness. Electron microscopy has been used for quantitative, comparative assessments of diversity in natural communities as well, but the resolving power cannot be fully exploited in this context for technical and economic reasons. Negative staining, which reveals the greatest detail of virions, is difficult to control and best suited to qualitative rather than quantitative assessments of diversity. Even with the greater uniformity provided by positive staining, it is difficult to confidently discriminate similar viruses in a mixed assemblage. A single virus may appear different depending on its orientation or as a result of damage or distortion during preparation. As a consequence of these limitations, quantitative surveys of morphological diversity have relied on classification into a handful of broad categories (e.g., <30, 30 to <60, 60 to <80, 80 to <100, and Z100 nm). The results of TEM surveys indicate virus-like particles in the ocean are usually numerically dominated by those with capsid diameters in the range of 30–60 nm, but range in size from around 20 to 750 nm. Viruses in the ocean thus span a similar size range as that observed among all classified viruses (18–500 nm). A significant, but variable, proportion of the viruses in seawater enumerated by TEM have tails, suggesting that they belong to the order Caudovirales and infect bacteria or archaea.

Genome Size

Genomes of viruses vary over orders of magnitude in length. Among marine viruses, for example they range from 3000 nt to 1,290,000 bp. Quantitative analysis of the genome length distribution in a sample can therefore provide some sense of diversity. Because conventional electrophoresis can only resolve DNA if it is less than about 20,000– 30,000 bp in length, pulsed-field gel electrophoresis (PFGE) has been used to analyze natural assemblages of viruses. The size range of virus genomes detected in seawater with this technique ranges from around 20,000 to 4500,000 bp. Under the right conditions, complex, fine-scale banding patterns can be seen, indicating that natural viral communities are quite diverse, but not all genome sizes are equally represented. In a wide variety of marine environments, the distribution of genome sizes appears to be multimodal, with peaks of genome copy numbers frequently occurring for sizes in the vicinity of 30, 55, and 60 kb. The peaks are broad, however, and there are bands detectable throughout the range. Bands 4100 kb are usually detected, meaning they contain a significant amount of DNA, but because there is more DNA per large genome, the copy number is low. The mean genome size of dsDNA viruses in seawater is around 50,000 bp, but the mode is closer to 30,000 bp. Because much of the DNA is clustered into several narrow size ranges, many genomes are not adequately resolved by PFGE. This method is therefore of limited value in quantifying viral diversity, especially richness, but remains a convenient means to track changes in com- munity composition.

Single-gene Surveys

Investigation of the diversity of microbial communities was revolutionized with the development of methods for cultivation-independent molecular surveys. Development of the polymerase chain reaction was transformative because it provided a means to amplify only specific genes of interest from a complex background of DNA. Analysis of the amplified material by sequencing or molecular fingerprinting allows one to determine how many variants of the targeted gene were present and thus evaluate genetic diversity. The small subunit ribosomal RNA gene has been the primary target for analyzing the diversity of cellular life, because of its many desirable properties including the fact that it is present in all cellular organisms. Viruses, however, do not code for ribosomal proteins. Viruses are so diverse, in fact, that there is no single gene that can be used as a universal viral target. Nevertheless, one can identify genes that are present and sufficiently conserved within subgroups of viruses that they can be used to survey diversity for a portion of the viral community. For example, structural genes of myovirids and phycodnavirids, DNA polymerase genes of podovirids and phycodnavirids, and the RNA polymerase gene of picornavirad have been used to assess the diversity of these various groups in the marine environment. Results from these surveys indicate that members of those viral groups are persistent and globally distributed, but within each group the specific sequence types detected and their relative abundance varies as a function of location, time, and depth. Comparing the viral gene sequences recovered by this cultivation-independent technique with homologous genes from viruses in culture has been very revealing. We now know that most of the viral diversity present in seawater is not represented in culture collections.

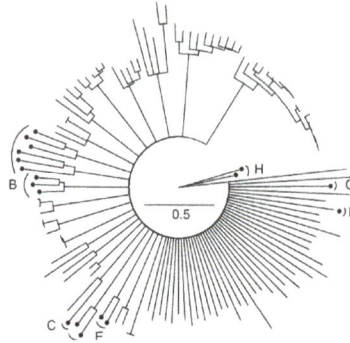

Comparison of the diversity among cultivated and uncultivated viruses

The extensive diversity of viruses revealed by single-gene surveys and how it compares to the diversity among viruses that have been classified is illustrated using marine picorna- like viruses as an example. A phylogenetic tree was created using partial sequences of the RNA-dependent RNA polymerase gene for picornavirad strains in culture (branches marked with a black circle at the tip) or those identified from single-gene surveys of seawater samples. Groups of sequences identified by an arc and a letter contain representative members of established groups of picornavirads as follows: group A, Secoviridae; group B, Bacillarnavirus; group C, Labyrnaviridae; group D, Picornaviridae; group E, Dicistroviridae; group F, Marnaviridae; group G, Iflaviridae; and the outgroup H, Potyviridae.

Metagenomics

There are a number of limitations implicit in the targeted gene approach. First, one recovers information only about variants of a single gene (usually just a portion of the gene). Second, those genes come from only a subset of the viral community, and third, creating a meaningful target in the first place re- quires at least some prior knowledge about the viruses being sought. Metagenomic analysis, in contrast, allows for a relatively unbiased sampling of all of the viral genomes present in a sample, with no prior knowledge of their sequences. To analyze a viral metagenome, viruses are first harvested from a seawater sample, separated from cells and dissolved nucleic acids, then the viral nucleic acids are extracted. The extracted genomic material is then broken into random, small pieces for sequence analysis. For technical reasons, metagenomic analyses have so far tended to focus on either the DNA viruses or the RNA viruses present in a sample, but it is possible to analyze both at once.

After the analysis, one has a collection of sequence reads that derive from random fragments of the various viral genomes that were present in the original sample. Because one is sampling diversity across entire genomes, rather than a single gene, metagenomic analyses require much greater numbers of sequencing reads (thousands to millions) than are typical of single-gene surveys (tens to hundreds) to draw meaningful conclusions about diversity within a sample and how it compares to other samples. A first-order analysis one can perform is to compare each of the sequences that make up a metagenome to all of the sequences archived in various repositories to see if they are similar to genes that are already known. One can then classify each sequence according to its likely function and the type of organisms from which it might derive. In addition to analyzing each read independently, it is also informative to determine how many of the sequence reads derive from overlapping regions of the same genome, and can thus be reassembled into a single longer read. The higher the diversity in a sample, the lower is the probability that any two or more reads in the randomly selected pool of sequences that comprise the metagenome will overlap. By making a number of assumptions, one can estimate the diversity of the original sample from the assembly statistics for the metagenome. Although metagenomics has its limitations, it is providing many new insights into the genetic diversity of viruses in the environment.

DNA-Containing Viruses

In metagenome analyses of marine DNA-containing viruses, most (around 75%) of the viral genes do not resemble any gene with a known function or which derive from a known organism. Among those sequences that are recognizable, the largest proportion (90–95%) of the recognizable reads is most similar to genes of bacteriophages in the order Caudovirales. A much smaller proportion resembles ssDNA-containing phages and sequences from eukaryotic viruses, primarily those in the families Phycodnaviridae and Mimiviridae. Since these data are presumed to represent a random sampling, this suggests that the most common DNA-containing viruses in the ocean are viruses that infect bacteria, followed by viruses that infect protists. DNA-containing animal viruses are very rare in the virioplankton, but metagenomic analyses of samples taken directly from marine animals do reveal sequences resembling known animal viruses.

The percentage of viral DNA metagenomic sequences that can be reassembled into larger fragments is typically low. Statistical analysis of assembly data suggest that there are many thousands of distinct virus types in a cubic meter of coastal seawater and tens of thousands in a kg of coastal sediment. The shape of the rank-abundance curves for virus sequence types appears to be best described by a power-law function. Even the most abundant sequence type in these communities is estimated to make up only around 1% (water) or0o.01% (sediment) of the total, which illustrates the exceptional diversity of marine viral communities.

RNA-Containing Viruses

The results from analyses of metagenomes prepared from viral RNA have been quite different than those prepared with viral DNA. A larger proportion of the reads (around 50%) are similar to known genes and almost all of them resemble eukaryote-infecting viruses. A small percentage of reads appear to derive from ribosomal RNA from cellular contamination, but none of the sequences resemble any of the known RNA bacteriophages. Most of the reads recognizable as viruses are most similar to sequences deriving from ssRNA-containing viruses in the order Picornavirales, but sequences resembling ssRNA viruses in other families outside of that order such as the Tombusviridae and dsRNA-containing reoviruses have also been detected.

Larger percentages of the sequences in RNA viral meta- genomes have been assembled than has been the case for DNA metagenomes. In fact, entire genomes have been readily assembled from random reads. Although RNA viruses are smaller than DNA genomes on average, the results also suggest that RNA viral diversity in the sea may be lower than that of the DNA viruses. However, there are no reliable estimates yet of just how diverse the RNA virus community might be and few environments have been sampled so far.

Marine Eukaryotic

Marine Fungi

It is important to consider that fungi evolved in the oceans in order to assess the diversity of marine fungi. While opinions differ as to when exactly fungi evolved from their ancestors, it is now generally accepted that the event took place at least a billion years before present. According to Heckman, this event probably took place around 1500 million years ago (Ma). The earliest plants, the chlorophycean algae appear to have emerged1100 Ma. It is also believed that plants conquered land much earlier than thought before, roughly around 1000 Ma, with the help of symbiotic fungi. Even if the exact dates of fungal origin are debated, fungi doubtlessly originated in the sea. This would mean that the lineages of early true marine fungi, which evolved in the sea prior to their conquest of land together with plants, probably continue to exist in the sea. As on A.D. 2000, about 450 species of obligate mycenean marine fungi have been described. Not many species have been described since then. Among the truly obligate marine fungi of the Kingdom Mycenae known at present, those belonging to the Ascomycotina and mitosporic fungi are the most common. Basidiomycetes and the earliest of true fungi, the chytrids, are few in number, the Zygomycotina being almost absent. When compared to nearly 70,000 species of fungi known so far on earth (out of 1.5 million estimated total number, a factor of ca.), the number of approximately 700 species of marine, mycenaean fungi is meagre. Even if the total number of obligately marine Mycenaean fungi is estimated to be 14,000 (20 times that of 700), this represents a disappointingly poor diversity of fungi in the sea compared to land, casting even serious doubts about their importance in the marine ecosystem. The straminipilan fungi belonging to Labyrinthulomycetes, which are truly marine, are more frequent in the oceans than the mycenean fungi, but even here the known number of species does not exceed about 50. The most 'primitive' of mycenean fungi, the chytridiomycetes are also extremely rare in the marine environment.

Two modern views of eukaryote classification

Cavalier-Smith[10]	Adl et al.
Kingdom Protozoa	Supergroup Amoebozoa Supergroup Rhizaria Supergroup Excavata
Kingdom Fungi (Kingdom Mycenae in this paper) Kingdom Animalia	Supergroup Opisthokonta First Rank Fungi (Mycenae in this paper) First Rank Mesomycetozoa First Rank Choanomonoda First Rank Metazoa
Kingdom Plantae Kingdom Chromista Subkingdom Cryptista Subkingdom Chromobiota Infrakingdom Haptista	Supergroup Archaeplastida Supergroup Chromalveolata Rank Cryptophyceae First Rank Haptophyta First Rank

Infrakingdom Heterokonta (containing the Hyphochytriales, Labyrinthulomycetes and Peronosporomycetes)	Stramenopiles (containing the fungal groups Hyphochytriales, Labyrinthulomycetes and Peronosporomycetes) First Rank Alveolata

Thraustochytrids are unicellular, produce plasma membrane extensions, the ectoplasmic net elements, and reproduce by means of zoospores. As is typical of the Kingdom Straminipila, these zoospores possess two unequal (heterokont) flagella, the longer anterior flagellum bearing tripartite hairs. They are also characterized by mitochondria with tubular cristae. Aplanochytrids may be unicellular or colonial and reproduce by spores that glide using ectoplasmic net elements, or by zoospores as above. Labyrinthulids are colonial and move within ectoplasmic net elements. Despite the fact that the first thraustochytrid was described by the great mycologist F.K. Sparrow nearly 70 years ago and that labyrinthulids have been known for a long time, only about 50 species of Labyrinthulomycetes are known so far. These fungi can be detected using the acriflavine direct detection (AfDD) method and cultured from natural samples using simple organic nutrient media. Several studies in recent years have demonstrated high abundance of Labyrinthulomycetes in the water column, using the AfDD method. We enumerated Labyrinthulomycetes in the water column of the Arabian Sea, using both the AfDD method, as well as culture methods. Cultured thraustochytrids amounted to only 0.06 to 3.65 % of the total thraustochytrids. In general, we have met with very little success in culturing thraustochytrids from oceanic waters.

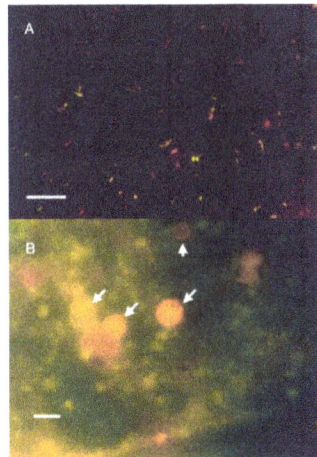

Difference between True Marine Fungi and Terrestrial Fungi

Lopez-Garcia examined the molecular diversity of eukaryotes within the range of 0.2 – 5 μm from the aphotic zone of 250 – 3000 m of the Atlantic polar front. Moon-vander Staay, likewise, studied picoplankton below the size fraction of 3 μm at a depth of 75 m in the equatorial Pacific. Their studies, using 18S rRDNA signatures,revealed the startling fact that fungal signatures were present in picoplankton samples.

The discovery of picoplanktonic, photoautotrophic cyanobacteria was discovered more than 50 years ago by Butcher in 1952 and was further confirmed by studies using microscopy and pigment composition. Further studies on the presence and importance of these phytoplankton of bacterial size, belonging to species of Synechococcus in 1979 and the discovery of species of Prochlorococcus in 1988 heralded research on picoplanktonic eukaryotes. Earlier studies on prokaryotic, autorophic picoplankton also revealed a complex assemblage of phototrophic eukaryotes of < 3 μm. Some of the small eukaryotic photosynthesizers were subsequently described as members of a new order, Parmales. The smallest eukaryote ever found, Ostreococcus tauri, is a member of the Prasinophyceae and has a cell size of 0.8–1.1 μm × 0.5–0.7 μm. It contains a compact, tightly packed cellular organization comprising the nucleus, one mitochondrion and one chloroplast tightly. Such a cellular organization also appears in other photosynthetic picoeukaryotes such as the Bolidophyceae.

It was not till the two publications that heterotrophic eukaryotes of the picoplankton size were discovered. Lopez-Garcia found signatures of picoplanktonic true fungi at a depth of 3000 m in the Antarctic waters, picoplanktonic thraustochytrids at 2000 m and a new lineage of heterokonts, close to Oomycetes, at 250 m. Moon van- der Staay discovered molecular signatures of eukaryotic Oomycetes and an early heterotrophic divergence of stramenopiles in the Pacific waters. Significantly, the signatures observed in these waters revealed what appeared to be novel picoplanktonic organisms belonging not only to fungi, but also to the alveolates (ciliates and dinoflagellates) and heterokonts (stramenipilan organisms, including stramenipilan fungi). Among fungi, these authors discovered a novel group of sequences at the base of fungi and among the stramenipilan fungi, three phylotypes distantly related to thraustochytrids and labyrinthulids. Scanning electron micrographs of picoeukaryotes from 200 m deep Mediterranean waters showed cells in the range of 1 μm or less, one of these with heterokont flagellae, as in the stramenopiles.

It now appears possible that picoplanktoic single- celled fungi may be prevalent in the sea. If so, a fresh look at our conventional notions of how a fungus looks may need revision. It is also possible that marine single-celled mycenaean fungi have so far been placed wrongly as protozoans. Several members of choanozoans are a case in point. Choanozoans occupy a crucial position in understanding the origin of animals and fungi from a common ancestor. It has recently been shown that at least some of the members of choanozoans may actually lie at the base of fungal evolution. Thus, the single-celled nucleariids have been shown to be a sister clade to mycenean fungi. The single-celled fish parasites, the ichthyosporeans too are probably similarly related. The single-celled choanozoan protist, Corallochytrium limacisporum Raghukumar, actually possesses the fungal signatures ergosterol and the α-aminoreductase gene of the lysis synthesis pathway, thus displaying a relationship to fungi.

The mycelial form of fungi is suited to penetrate particulate organic matter and traverse space filled with air. Such a form is unsuitable for organisms living planktonically in the water column, a habitat that favours single-celled forms more. Therefore, it may well turn out that fungi at the base of the clade of the Kingdom Mycenae are single-celled and not mycelial and are typical inhabitants of the sea. A search for such organisms in the future may actually enhance our knowledge on the diversity of marine fungi.

Fungi in Extreme Marine Environments

Fungi have never been considered to be truly anaerobic in respiration. The only obligate fungal anaerobes known so far belong to the Neocallimasticales belonging to chytrids. No other such fungi have been discovered so far. Could anaerobic fungi be prevalent in the sea? The work of Dawson &

Pace using 18S rRDNA gene diversity in intertidal, black, anoxic sedments has revealed the extensive presence of eukaryotes in these habitats. These authors discovered numerous novel branches among the stramenopiles, including the fungal groups of oomycetes, hyphochytriomycetes and labyrinthulo- mycetes, as well as among protozoans. They further suggested that anaerobic eukaryotes may be the phylogenetically most diverse organisms in the sea. Likewise, Stoeck & Epstein studied the diversity of micro-eukaryotes in the suboxic waters and anoxic sediments of salt marshes. Most of the sequences represented deep, novel branches within green plants, cercozoans, and alveolates, as well as within mycenaen and stramenopilan fungi. Further work of Stoeck. on the permanently anoxic waters of the Cariaco Basin in the Caribbean Sea revealed organisms belonging to deep branches within the above groups and three novel lineages branching at the base of the eukaryotic evolutionary tree. These preceded, were contemporary with, or immediately followed the earliest eukaryotic branches. A significant number of these appeared to be new organisms. They suggested that these newly discovered protists might retain traits reminiscent of an early eukaryotic ancestor, at the base of the evolutionary trees. These exciting papers have a tremendous implication to marine fungal diversity.

Two of the exciting locations for investigations on anaerobic fungi in the Arabian Sea are the permanent oxygen minimum zone (OMZ) of the northern and central Arabian Sea and the seasonal OMZ of the west coast of India. It has been well known that the former is characterized by high densities of denitrifying bacteria. Jayakumar examined the diversity in the Arabian Sea OMZ of the prokaryotic 132 nirS gene sequences involved in the conversion of nitrite to nitric oxide. These authors found 12 major clusters, most of which did not show a high level of identity with other nirS sequences reported earlier. The dominant type in one of the surface samples was close to Pseudomonas aeruginosa. Highest diversity was found in samples with high nitrite. It is likely that fungi characterized by anaerobic denitrification also occur in such regions. Thraustochytrids are present in high numbers at these OMZ depths and their physiology is likely to provide us some surprise. Indeed, recent reports found several terrestrial fungi capable of converting nitrate to ammonia through denitrifying processes. Preliminary work carried out on such fungi in the coastal waters of Goa has already yielded several fungi, albeit belonging to known species, which are capable of denitrification.

Exotic Habitats of Marine Fungal Diversity

Several exotic marine ecosystems are yet to be studied intensively for fungal and other microbial diversity. Coral reefs, the deep sea and hydrothermal vents are some of the examples. Few obligate marine fungi have been reported from these habitats. For example, the Kohlmeyers described several marine fungi from dead corals. Surprisingly, not many have been added to this list since then, contrary to that of lignicolous fungi. Lignicolous marine fungi presently comprise about 450 species, new species having been discovered consistently since the seminal paper of Barghoorn & Linder in 1944.

Bacteria associated with invertebrates; especially in coral reef habitats is an exciting new area, of particular interest in the discovery of bioactive molecules. Although several fungi associated with sponges and producing interesting new molecules have been reported, all such species belong to terrestrial genera and few obligate marine fungi associated with such animals have been reported so far. This is a fertile area for future research. Similar is the case with deep-sea fungi. Terrestrial species of fungi adapted to deep-sea conditions probably play an important role in the ecology of deep-sea sediments. Fungal mycelia are easily detected in such sediments and fungi are probably prevalent in these habitats. Kohlmeyer & Kohlmeyer described obligate deep- sea fungi from decaying wood and bryozoan shells. Few have been described since then.

The evolutionary history of fungi strongly suggests that obligate marine fungi, unlike those reported so far from the sea are bound to exist. Biodiversity-rich ecosystems such as the coral reefs, the deep-sea and anoxic sediments are bound to harbour a high diversity of obligate marine fungi. Unfortunately few have been described so far. The paucity in our knowledge on the real diversity of obligate marine fungi in the above habitats brings us back to the questions that we asked in the beginning and we may conclude as follows.

1. We might have failed to recognize obligate marine fungi in exotic habitats because they look different from our conventional wisdom of fungi. Corallochytrium limacisporum Raghukumar, an inhabitant of coral reefs waters, the nucleariids and the ichthyosporeans discussed earlier are possible examples of atypical fungi in the marine environment.

2. Culture methods for marine fungi have been highly inadequate so far.

3. Molecular methods will provide us greater insights into marine fungal diversity.

Marine Protists

Protists are microscopic eukaryotic microbes that are ubiquitous, diverse, and major participants in oceanic food webs and in marine biogeochemical cycles. The study and characterization of protists has a long and distinguished tradition. Even with this history, the extraordinary species diversity and variety of interactions of protists in the sea are only now being fully appreciated. Figure below shows representative examples of marine protists, and of methods used to visualize these microbes.

Protists can be autotrophic or heterotrophic. The former, also called microscopic algae, contain chloroplasts, thrive by photosynthesis, and are at the base of all oceanic food webs, with the exception of deep-sea chemosynthetic ecosystems. There is a general trend for > 20-μm sized phytoplankton (microplankton), such as diatoms and dinoflagellates, to dominate episodically in coastal waters, while in the open ocean 2–20-μm sized cells (nanoplankton), such as coccolithophorids, sporadically form massive blooms that can be seen from space. However, during nonbloom seasons, even smaller cells (picoplankton— prokaryotes and eukaryotes less than a few micrometers in size) typically dominate phytoplankton biomass and production. Some genera of marine picoalgae can "bloom" to very high concentrations (> 105 cells ml-1), Ostreococcus, while others are ubiquitous, such as Micromonas, which is found from arctic to tropical waters.

In contrast to photosynthetic protists, heterotrophic protists have no permanent chloroplasts and rely on other organisms for nutrition. Most are phagotrophic— they ingest prey—usually other microbes. Some heterotrophic protists are parasites of phytoplankton or zooplankton, while still others form symbiotic relationships with autotrophic or heterotrophic bacteria. In addition, many species of protists have mixed trophic modes. These "mixotrophs" include flagellated phytoplankton that ingest bacteria or other protists, ciliates that "farm" chloroplasts from ingested algal prey, and mutualistic relationships between photosynthetic microorganisms and heterotrophic protists. Heterotrophic nutrition occurs among virtually all lineages of protists. Some protistan groups formerly classified as "algae" include species with strictly heterotrophic nutrition. For example, the dinoflagellates include many heterotrophic species, and it is possible that all dinoflagellates, including the photosynthetic species responsible for red tides, may have the capacity for phagotrophy.

Phagotrophic protists have long been known to be important in oceanic food webs as consumers of bacteria and phytoplankton, as regenerators of nutrients for further phytoplankton growth, and as a food resource for marine zooplankton. The grazing impact of phagotrophic protists has consequences for both ecosystem modeling and for the structure of microbial communities. The proportion of organic carbon produced by phytoplankton that flows through a multi-step microbial food web versus a shorter phytoplankton- mesozooplankton food chain is a critical factor that determines the capacity of marine ecosystems to sequester organic carbon and to efficiently produce fish biomass. Via selective grazing, both herbivorous and bacterivorous protists can alter the community composition of oceanic phytoplankton and bacterial assemblages. It is clear that these activities constitute major roles in oceanic ecosystems, and considerable information has been amassed on the abundances, distributions, and ecological roles of protists, yet there is still much to learn. At present, much of our knowledge is restricted to larger species that have readily identifiable morphological traits and those species that are amenable to laboratory culture.

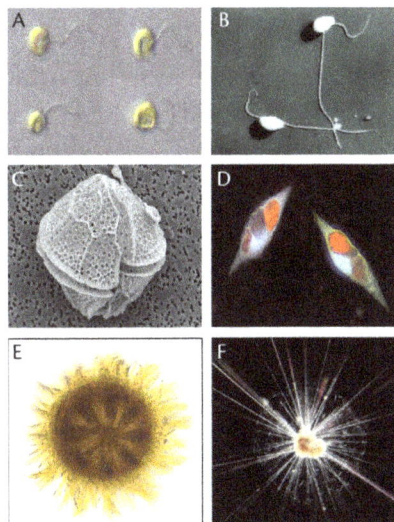

Examples of marine protists and of methods
used to visualize eukaryotic microbes.

(A) Live cells of the 2-µm-diameter marine flagellated alga Pelagomonas calceolate, transmitted light microscopy. (B) Two cells of a 5-µm-long marine heterotrophic flagellate (Bodo sp.), scanning electron microscopy (SEM). (C) A 50-µm-diameter autotrophic thecate dinoflagellate (Lingulodinium polyedrum) that forms red tide blooms, SEM. (D) Two cells of a 20 x 60 µm marine heterotrophic dinoflagellate (Gyrodinium sp.) with food vacuoles full of ingested coccoid cyanobacteria, epifluorescence microscopy after preservation with formaldehyde and staining with the fluorochrome DAPI. (E) The 55-µm-diameter marine pelagic ciliate Strombilidium sp., inverted light microscopy after preservation with acid Lugol fixative. (F) Live foraminiferan Hastigerina pelagica with a fluid bubble capsule and a test approximately 300 µm across, darkfield light microscopy.

One fundamental question that has become a focus of research and debate in recent years revolves around the species concept (and species boundaries) for protists. Although protistan species have traditionally been defined based on their morphology, recent molecular analyses and physiological studies reveal that even well defined morphospecies may be composed of a mosaic of multiple genetic/physiological types. For instance, the ubiquitous diatom Skeletonema costatum, thought until recently to be a single species, is actually a composite of ten genetic types that are now considered distinct species. Similarly, physiological variability among Spumellalike heterotrophic flagellates indicates geographically distinct adaptations and emphasizes that morphology needs to be complemented by other approaches for distinguishing different species. The situation is even more complicated for the tiny picoplanktonic protists in the ocean. Molecular analyses during the last ten years have revealed many undescribed and uncultivated taxa among these morphologically similar forms, and whole lineages in the alveolate and stramenopile divisions have escaped detection until recently. The discovery of many cryptic lineages highlights the importance of combining molecular and ecological information with traditional morphological descriptions. This combination facilitates meaningful investigation of the distribution and activities of marine protists in situ.

These new approaches for defining protistan species have energized an ongoing debate regarding protistan species diversity and biogeography. The overarching question is whether there are relatively few species of protists that are broadly distributed, or whether similar protistan morphotypes are in fact distinct species or subspecies with more limited distribution. Why does species diversity matter? From an ecological perspective, it is important to know what taxonomic level (e.g., species or subspecies) yields information about associated differences in functional roles (i.e., how these organisms participate in food webs). Based on analysis of genetic variation, Katz found evidence for a "cosmopolitan" distribution (high gene flow and low diversity) for species of ciliates in coastal waters and evidence for "endemism" (high diversity and geographically restricted gene flow) among species of ciliates in isolated tide pools. Little is known at this juncture about the distribution, gene flow, diversity, ecological roles, or even the morphologies of the previously undescribed lineages of protists now being revealed by molecular analyses.

Beyond characterizing the diversity and distribution of protists in the ocean, major lines of research continue to elucidate the ecological roles of protists in marine ecosystems. Species-specific growth, grazing, and nutrient excretion rates of heterotrophic protists are derived mainly from laboratory experiments using isolated species fed monospecific prey under conditions that poorly mimic natural systems. However, such studies show that phagotrophic protists exhibit various types of species- specific behavior that may affect feeding behavior and growth, including chemosensory responses to prey and selective ingestion of prey types. In turn, prey cells may use structural or chemical defenses against protistan predation. Research on the underlying physiological/biochemical basis of the feeding behavior of marine protists is yielding new awareness and understanding of their predator-prey interactions. The ability of herbivorous protists to discriminate between alternate prey types depending on size and "taste" undoubtedly affects protistan-grazing impact on bloom-forming phytoplankton, such as diatoms and harmful algal species. Studies of protistan herbivory on in situ phytoplankton communities indicate that food quality can play an important role in grazer selectivity and that protistan growth efficiencies vary greatly with prey type. Accurately describing the rates of activity of heterotrophic protists is extremely important to carbon-cycle models.

Unquestionably, many fascinating discoveries are ahead for protistan researchers at the physiological, organismal, and community levels. These include:

- characterizing the overall breadth and meaning of protistan species diversity in the ocean;

- understanding the ecology of minute organisms, such as the 2 μm heterotrophic flagellate Symbiomonas scintillans that itself harbors an endosymbiotic bacterium and the pico-alga Ostreococcus tauri, the smallest free-living eukaryote known, which can occur at high abundances and has unique genome features;

- deciphering the ecological, molecular, and biochemical processes that might explain chloroplast acquisition by heterotrophic protists; and

- characterizing microbial interactions and processes sufficiently to provide a predictive understanding of community function and how microbial communities will respond in the face of environmental change.

Yet another challenge ahead is to understand the importance of processes such as parasitism and symbiosis in regulating oceanic plankton community structure and production; both are known to occur, but little is known about their prevalence and extent in marine systems.

Breakthroughs will depend on the application of a diverse array of approaches and methodologies. These will almost assuredly hold implications even beyond the important roles these organisms are already known to play in oceanic ecosystems. Marine protists likely retain some characteristics of the earliest eukaryotes that evolved on Earth. Interrogation of their genomes provides insights into how they thrive in the world's ocean and insights into fundamental biological processes such as the evolution of multicellularity, and thus will foster a better understanding of the evolution of life on our planet.

Seaweed

Seaweed is the common name for marine algae—a group of species from the Protista kingdom, meaning they are not plants at all, even though they may look like underwater plants, growing to more than 150 feet in length.

Algae are not plants, although they do use chlorophyll for photosynthesis, and they do have plant-like cell walls. However, seaweeds have no root system or internal vascular systems; nor do they have seeds or flowers.

Types of Marine Algae

Marine algae are divided into three groups:

>   Brown Algae (Phaeophyta)
>
>   Green Algae (Chlorophyta)
>
>   Red Algae (Rhodophyta)

There is a fourth type of algae, the tuft-forming bluegreen algae (Cyanobacteria) that is sometimes considered to be seaweed.

Brown Algae: Phaeophyta

Brown algae is the largest type of seaweed. Brown algae is in the phylum Phaeophyta, which means "dusky plants." Brown algae is brown or yellow-brown in color and found in temperate or arctic waters. Brown algae typically have a root-like structure called a "holdfast" to anchor the algae to a surface.

One type of brown algae forms the giant kelp forests near the California coat, while another forms the floating kelp beds in the Sargasso sea. Many of the edible seawoods are kelps.

Examples of brown algae: kelp, rockweed (Fucus), Sargassum.

Red Algae: Rhodophyta

There are more than 6,000 species of red algae. Red algae has its often brilliant color due to the pigment phycoerythrin. This algae can live at greater depths than brown and green algae because it absorbs blue light. Coralline algae, a subgroup of red algae, is important in the formation of coral reefs.

Several types of red algae are used in food additives, and some are regular parts of Asian cuisine.

Example of red algae: Irish moss, coralline algae, dulse (Palmaria palmata).

Green Algae: Chlorophyta

There are more than 4,000 species of green algae. Green algae may be found in marine or freshwater habitats, and some even thrive in moist soils. These algae come in three forms: unicellular, colonial or multicellular.

Examples of green algae: sea lettuce (Ulva sp.), which is commonly found in tidal pools, and Codium sp., one species of which is commonly called "dead man's fingers."

Structure of Seaweeds

Thallus: the entire body of seaweed.

Lamina: a flattened structure that is resembles a leaf.

Sorus: a cluster of spores spore.

Air bladders: a hollow, gas-filled structure organ which helps the seaweed float, found on the blade). Other seaweeds (e.g. kelp) have floats which are located between the lamina and stipe.

Stipe: a stem-like structure, not all seaweeds have these.

Holdfast: a specialized structure on the base of seaweed, which acts as an "anchor" allowing it to attach to a surface (e.g. a rock).

Haptera: finger-like extensions of holdfast anchoring to benthic substrate.

Seaweeds play very important roles in many marine communities. They are a food source for many marine animals such as sea urchins and fishes, and form the base of some food webs. They also provide shelter and a home for numerous fishes, invertebrates, birds, and mammals.

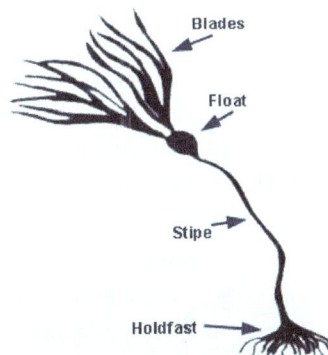

Structure of marine algae

Seaweed Reproduction

Seaweed life and reproductive cycles can be quite complicated. Some seaweeds are perennial, living for many years, while are annuals. Annual seaweeds generally begin to grow in the spring, and continue throughout the summer. Some red seaweed has a life span of 6 to 10 years.

Seaweeds can reproduce sexually, by the joining of specialized male and female reproductive cells, called gametes. After they are released from the sporophyte, the spores settle and grow into male and female plants called gametophytes. The gametophytes produce gametes (sperm or eggs). The sperm and eggs are either retained within the gametophyte plant body, or released into the water. Eggs are fertilized when the sperm and egg fuse together, and a zygote is formed. Zygotes develop and grow into sporophytes, and the life cycle continues.

Seaweeds display a variety of different reproductive and life cycles and the description above is only a general example of one type, called alternation of generations. In a few species there is an alternating sexual and asexual reproductive process with every generation.

Seaweeds can also reproduce asexually through fragmentation or division. This occurs when parts of a plant break off and develop directly into new individuals. All offspring resulting from asexual reproduction are clones; they are genetically identical to each other and the parent seaweed.

Uses of Seaweed

Seaweeds are food source for humans especially in East Asia, it is most commonly associated with Japanese food. Seaweeds also are used to make a number of food additives such as alginates and carrageenan which is used in cooking and baking as a vegetarian alternative to gelatine.

Carrageean products

Many seaweeds are used as medicine. Alginates are used in wound dressings and in the production of dental moulds and agar is used very widely in Microbiology to help grow bacterial cultures.

Agar plates

Seaweeds are ingredients in toothpaste, cosmetics and paints and are used in industrial products such as paper coatings, adhesives, dyes, gels, explosives and many more.

Japanese food uses seaweeds extensively - Kombu
a brown alga and Kim nori a red alga

Nori

Much of the oil and natural gas we use today formed from seaweeds which partially decomposed on the sea floor many millions of years ago.

Red Seaweed

Prokaryotes

Marine Bacteria

Marine bacteria are single-celled organisms that can be shaped like little spheres, rods, or (less commonly) spirals. They are often very small, with cell diameters of just a few microns (about 1/100th the width of a human hair). They perform all kinds of chemical processes in the open ocean, including most of the steps in nitrogen cycling.

Cyanobacteria are a large group of photosynthetic bacteria, some of which can "fix" nitrogen, converting nitrogen gas into more biologically useful compounds. Cyanobacteria live in all kinds of environments, but are especially important in open-ocean ecosystems. They were formerly known as "blue-green algae" but are now recognized as a type of bacteria, not a type of algae (algae are eukaryotes).

Marine Bacteria that Fixes Nitrogen

Trichodesmium

Trichodesimium is a genus of colonial cyanobacteria that is one of the most important and well-studied nitrogen-fixing organisms found in open-ocean areas such as Station ALOHA. It is one of the few organisms involved in the oceanic nitrogen cycle that is visible to the naked eye. Trichodesmium is a colonial organism that forms hair-like strands, which sometimes aggregate into tiny "puffballs" up to a millimeter or two across. When winds are light, Trichodesmium colonies may clump together and float right on the sea surface, where they are known as "sea sawdust."

Trichodesmium uses an enzyme called "nitrogenase" to transform nitrogen gas into more biologically useful compounds (a process called "nitrogen fixation.") However, nitrogenase is inactivated in the presence of oxygen (which is produced by many photosynthetic organisms as a byproduct of photosynthesis). Because of this, most nitrogen-fixing microbes separate the processes of nitrogen fixation and photosynthesis either spatially (using different types of cells for each process), or temporally, by performing photosynthesis in the daytime and fixing nitrogen at night.

Trichodesmium, however, uses neither of these strategies. It is able to fix nitrogen during the daytime, but does not have specialized cells to perform the job. Researchers are very interested in figuring out how Trichodesmium is able to fix nitrogen in the daytime.

During the BioLINCS cruise, Trichodesmium is being studied using microscopy, nitrogen-uptake experiments, and nucleic acid (DNA and RNA) analyses. These analyses will be carried both out on the ship and later on shore. Among other things, researchers will be using genetic "probes" that look for a gene called nifH, which is one of several genes that help microbes produce the nitrogenase enzyme. This will help them determine the abundance and nitrogen-fixing ability of Trichodesmium. The Environmental Sample Processor (ESP) will also be using the nifH gene to monitor the presence of Trichodesmium in real time.

Heterocystus Cyanobacteria

Heterocystus cyanobacteria are multi-celled organisms that form microscopic filaments and perform nitrogen fixation in the open ocean. The most common genus of heterocystus cyanobacteria in open-ocean areas is Richelia, which is almost always found living inside of diatoms, a type of microscopic marine algae.

As part of this "symbiotic" living arrangement, the heterocystus cyanobacteria provide the diatoms with "fixed" nitrogen and other nutrients. It is not entirely clear what the bacteria get out of the arrangement.

Like all nitrogen fixing organisms, heterocystus cyanobacteria use an enzyme called "nitrogenase" to transform nitrogen gas into more useable forms of nitrogen. However, nitrogenase is inhibited in the oxygen-rich environment that exists inside most cells. As multi-cellular organisms, heterocystus cyanobacteria have evolved specialized cells called "heterocysts," which provide an environment more conducive to nitrogen fixation.

During the BioLINCS cruise, heterocystus cyanobacteria are being studied using DNA analyses that are carried out manually on board the ship and performed automatically within the Environmental Sample Processor (ESP). Among other things, researchers will be looking for the nifH gene in Richelia diatoms. This is one of several genes that allow cyanobacteria to produce the nitrogenase enzyme. Researchers will also be combining DNA analyses with measurements using influx flow cytometer (IFC) to determine whether heterocystus cyanobacteria are associated with specific types of marine algae.

Crocosphaera

Crocosphaera is a genus of singled-celled cyanobacteria that lives through photosynthesis and can "fix" nitrogen. It grows in many tropical ocean areas where the water is about 24° Celsius (75° Fahrenheit). At about two to four microns across, Crocosphaera is much smaller than Trichodesmium. Like all nitrogen-fixing organisms, Crocosphaera uses an enzyme called "nitrogenase" to transform nitrogen gas into more biologically useful compounds. However, nitrogenase doesn't work well in the presence of oxygen (a byproduct of photosynthesis). As a solution to this problem, Crocosphaera performs photosynthesis during the day and nitrogen fixation at night.

Uncultivated Cyanobacteria Group A (UCYN-A)

UCYN-A is a group of nitrogen-fixing cyanobacteria that were first discovered in the open ocean near Hawaii in 1998. Most of the time, they are not the most abundant nitrogen-fixing organism, but they can reproduce rapidly when conditions are right.

UCYN-A bacteria contain very few genes, compared with most other marine cyanobacteria. Although the entire UCYN-A genome has been sequenced, the organisms themselves have yet to be grown in the laboratory.

Surprisingly, sequencing of the UCYN-A genome revealed that that the bacteria lack genes that are necessary for "fixing" inorganic carbon and for key parts of the photosynthetic process. This allows UCYN-A to fix nitrogen during the daytime. However, because the bacteria cannot perform photosynthesis, they must have some other means of acquiring key organic compounds that are normally produced during photosynthesis. As with some heterocystus cyanobacteria, it is possible that UCYN-A form "partnerships" (symbioses) with other photosynthetic organisms to acquire these essential compounds.

During the BioLINCS cruise, UCYN-A are being studied using DNA and RNA analyses that are carried out manually on board the ship and performed automatically within the Environmental Sample Processor (ESP).

Alphaproteobacteria and Gammaproteobacteria

Proteobacteria are an extremely diverse group of bacteria. Some alphaproteobacteria and gammaproteobacteria, two subgroups within the proteobaceria, carry the nifH gene, and may be able to fix nitrogen. The proteobacteria also include some key players in the process of "nitrification," as described below.

During the BioLINCS cruise, the Environmental Sample Processor (ESP) will be looking for the nitrogenase (nifH) gene specific to gammaproteobacteria. This gene allows the bacteria to produce the enzyme nitrogenase, which they use to "fix" nitrogen. The ESP will also be looking for alphaproteobacteria using sandwich hybridization arrays.

Marine Bacteria Involved in Nitrification

Ammonium Oxidizing Bacteria (AOB)

The ammonium oxidizing bacteria (AOB) are bacteria that are all involved in a specific biochemical process (nitrification), but which may or may not be related to one another. They are involved in the first step of nitrification—the conversion of ammonium to nitrite (also known as "ammonium oxidation").

Most of the known ammonium oxidizing bacteria are betaproteobacteria and gammaproteobacteria. These bacteria are believed to be involved in this process because they carry a gene that codes for the ammonia monooxygenase enzyme (amoA).

Nitrite Oxidizing Bacteria

Several different genera of marine bacteria are involved in in the second step in the nitrification process—converting nitrite to nitrate (also known as "nitrite oxidation"). These include Nitrobacter,

Nitrospira, and Nitrospina Of these, Nitrospira is believed to be the most important in the open ocean.

Marine Bacteria that are Primary Producers but do not Fix Nitrogen

Prochlorococcus

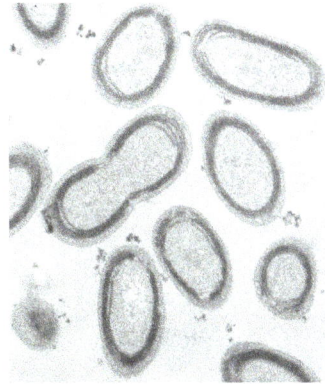

Prochlorococcus is a genus of cyanobacteria that is very common in open-ocean areas around the world. Although extremely tiny, with cells only 0.5 to 0.8 microns across, Prochlorococcus (along with another genus, Synechococcus) are so widespread and abundant that may produce a third of the oxygen in the Earth's atmosphere. Prochlorococcus are "obligate photoautotrophs," obtaining all of their energy through photosynthesis. However, they require nitrogen as a nutrient, and can use either nitrate or ammonium as a source of nitrogen. Thus, they are a key part of the "assimilation" process in oceanic nitrogen cycles.

During the BioLINCS cruise, researchers will be studying Prochlorococcus using the influx flow cytometer (IFC), and in some on-deck incubation experiments.

Synechococcus

Synechococcus is another common type of marine cyanobacteria. It is arguably the second most common group of photosynthetic marine bacteria, after Prochlorococcus. It has a similar shape as Prochlorococcus, but is typically a little bit larger, at 0.8 to 1.5 microns across.

Like Prochlorococcus, Synechococcus are "obligate phototrophs," which means that they can only obtain energy through photosynthesis, and require nitrate as a key nutrient. They can use nitrate or ammonium as sources of nitrogen, and are a key part of the "assimilation" process in oceanic nitrogen cycles.

During the BioLINCS cruise, researchers will be studying Synechococcus using the influx flow cytometer (IFC).

Marine Archaea

The Archaea are a curious phylogenetic domain (formerly kingdom) comprised of an odd assortment of cultured microbes that fall into three major groupings: extreme halophiles, methanogens, and extreme thermophiles and thermoacidophiles. Why such an odd assortment of salt-loving, or anaerobic, or heat-loving microbes should form such a coherent phylogenetic grouping is still not that well understood. The dogma until 1992 was that archaea inhabit mainly "extreme" environments, inhospitable to most other life forms. The existence of novel archaeal types was first hinted at during cultivation-independent ribosomal RNA surveys in open-ocean and coastal marine waters. Initial work used the polymerase chain reaction (PCR) to amplify ribosomal RNA genes from mixed microbial populations. In 1992, Jed Fuhrman of the University of Southern California first reported the existence of a new type of archaeal ribosomal RNA sequence from deepwater planktonic microbes in the Pacific Ocean.

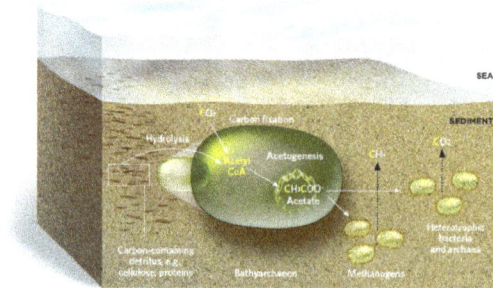

Archaea could be found in cold, aerobic habitats of coastal and open-ocean waters—and, to top it off, they were abundant. No cultivated, characterized archaea were known to grow at the combined salinity, temperature, and oxygen concentration found in temperate oceanic waters, shallow or deep. Following on the heels of these first oceanic sightings, archaeal groups began cropping up in many unexpected habitats. The next steps were to quantify the distribution and abundance of these unusual microbes and to begin to understand their biological properties and ecological significance.

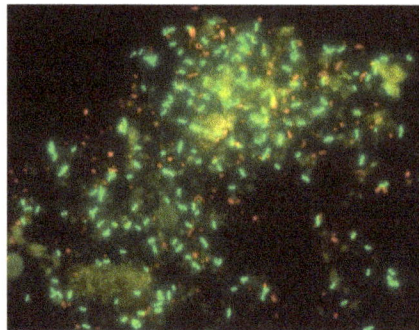

Marine Archaeal Abundance, Distribution, and Variability

Once it was clear that archaea were reasonably abundant players in microbial plankton, scientists initiated studies that employed radiolabeled, archaeaspecific oligonucleotide probes to quantify total, extractable archaeal rRNA in marine plankton. Surprisingly, planktonic Crenarchaea were found to contribute as much as 20% to the total microbial rRNA in late-winter Antarctic coastal waters at -1.8°C. Surveys in temperate waters off the coast of California also showed that the planktonic Crenarchaea tended to be most abundant in waters below the euphotic zone. This trend has generally held in a global sense in both coastal and open-ocean settings.

Ribosomal RNA-targeted fluorescence probes have been used to provide estimates of archaeal cell numbers in the water column. Off the California coast, for example, planktonic Crenarchaea represent > 20% of the total picoplankton cell counts from water depths of 80–3000 m. At the Hawaii Ocean Time-series (HOT) station ALOHA, Dave Karl and collaborators showed that Crenarchaea comprised as much as 30% of the total microbial counts in deep waters below the euphotic zone. In aggregate, these and other data suggest that pelagic Crenarchaea comprise a significant proportion of overall planktonic microbial biomass throughout the world's ocean.

Biology and Ecology of Planktonic Marine Crenarchaea

Creative application of biochemical, geochemical, and genomic techniques have provided consider-able data on the planktonic Crenarchaea. These studies, combined with isolation in pure culture of a marine crenarchaeon (see below), now provide some specific clues regarding the biogeochemical and ecological importance of this abundant marine microbial group. Lipid analyses of cold marine sediments by Jaap Damste's group at the Netherlands Institute for Sea Research (NIOZ) revealed a first for this environment—high levels of tetraether lipids, that were previously found only in ther-mophilic Crenarchaea. Stuart Wakeham at the Skidaway Institute of Marine Science and my group at the University of California, Santa Barbara, then showed that marine plankton samples with high numbers of planktonic Crenarchaea also contained high levels of the same tetraether lipids. In ad-dition, Preston showed that Cenarchaeum symbiosum, a crenarchaeal symbiont of marine sponges, contained the very same tetraether lipids. These could be used to infer the lipid's detailed chemical structure. Collectively, these data establish that the sources of abundant marine tetraether lipids in the plankton are indeed derived from planktonic Crenarchaea.

A real surprise came when Ann Pearson, during her graduate work with Tim Eglinton at Woods Hole Oceanographic Institution, provided radioisotopic data suggesting pelagic Crenarchaea may be chemoautotrophic. She purified large amounts of marine archaeal lipids from deep-sea surficial sediments derived from the deep-water planktonic Crenarchaea. Natural 14C isotope analyses of archaeal lipids suggested that deep-water archaea were not consuming much organic matter de-rived from surface primary productivity. Rather, deep-water Crenarchaea appeared to be using dissolved inorganic carbon as their main carbon source. Pearson's analyses are now supported by a variety of independent studies. For example, when Cornelia Wuchter, a graduate student at NIOZ, added 13C-labeled bicarbonate to a sample from the North Sea, then incubated the seawater in the dark, the heavy-isotope label from CO_2 was nearly exclusively incorporated into crenarchaeal lipids.

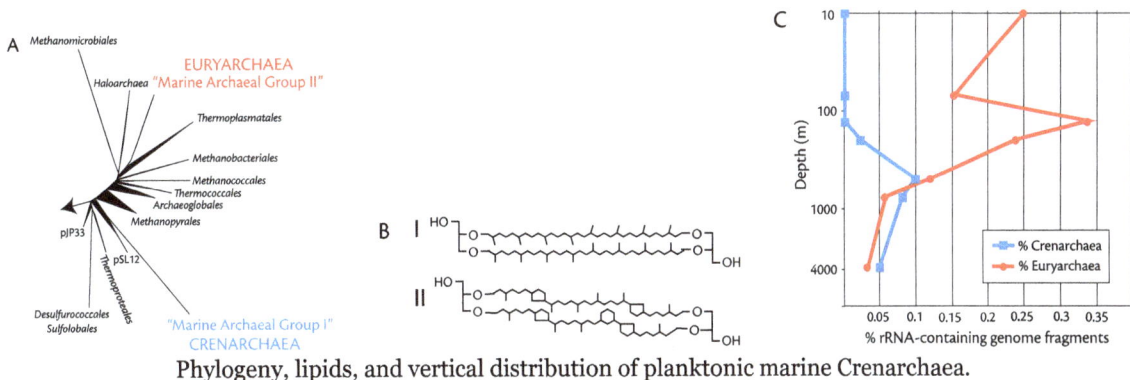

Phylogeny, lipids, and vertical distribution of planktonic marine Crenarchaea.

(A) A phylogenetic tree showing the general affiliations of two different planktonic archaeal groups.
(B) General structures of abundant tetraether lipids found in Crenarchaea. (C) Distribution of

planktonic Crenarchaea and Euryarchaea in the North Pacific Subtropical Gyre, as detected in large DNA-insert genomic libraries.

The observation of CO_2 -fixing Crenarchaea of course immediately led to the question: exactly what are these ubiquitous CO_2 -fixing marine Crenarchaea using as their energy source? Early clues came from the discovery of an ammonia monooxygenase gene in Crenarchaea that encodes a key enzyme used by nitrifiers in the first step of ammonia oxidation. This new archaeal ammonia monooxygenase gene was subsequently detected in many marine environments as well. Genomic analyses subsequently revealed many other genes associated with nitrification and CO_2 fixation in marine Crenarchaea. A definitive demonstration that marine Crenarchaea are indeed nitrifiers was achieved by Dave Stahl's group at the University of Washington, in collaboration with John Waterbury at the Woods Hole Oceanographic Institution. In cultures designed to isolate nitrifying bacteria (bacteria that oxidize ammonia obtain energy and utilize CO_2 as their carbon source), Stahl's group unexpectedly domain Bacteria. After some major microbe sleuthing, Stahl's group discovered that the new nitrifiers were, in fact, the same type of Crenarchaea that are so abundant in marine plankton. These crenarchaeal found ammonia-oxidizing cultures that did not appear to be typical bacterial nitrifiers—in fact, these nitrifying microbes did not even belong to the isolates grew exclusively on ammonia for energy, producing nitrite as the end product. The archaeal nitrifiers were also shown to use CO_2 as their carbon source, and they did not appear capable of growing on organic matter. These data, combined with earlier information on crenarchaeal distributions and abundance, indicate that planktonic Crenarchaea are indeed critical players in the ocean's nitrogen cycle, and they exert a large influence on oceanic nitrification in the sea.

New observations on the biology, ecology, and activity of planktonic Crenarchaea continue to accumulate. These include detailed quantitative analyses modeling in situ carbon sources of planktonic Crenarchaea, microautoradiography studies to track substrate assimilation into different cell types, and observations of Crenarchaea in anoxic zones of the Black Sea and the Arabian Sea. The potential interactions between Crenarchaea and other bacterioplankton have also recently been suggested by the similar distributions of bacterial nitrite oxidizers belonging to the genus Nitrospina and to Crenarchaea in the water column. Future studies promise to further elucidate the ecophysiological properties of the ubiquitous marine crenarchaeal nitrifiers and to better quantify and constrain nitrification and CO_2 fixation rates, as well as further characterize their ecological interactions.

Marine Microbial Diversity

Microbes were the only form of life for the first 2-3 billion years of planetary and biological evolution. Life most likely began in the oceans and marine microorganisms are the closest living descendants of the original forms of life. Early marine microorganisms also helped create the conditions under which subsequent life developed. More than two billion years ago, the generation of oxygen by photosynthetic marine microorganisms helped shape the chemical environment in which plants, animals, and all other life forms have evolved. Macroscopic life and planetary habitability completely depend upon the transformations mediated by complex microbial communities. These microscopic factories both aerobic and anaerobic are the essential catalysts for all of the chemical reactions within the biogeochemical cycles. Their unique metabolisms allow marine microbes to carry out many steps of the biogeochemical cycles that other organisms are unable to complete. The smooth functioning of these cycles is necessary for life to continue on earth.

Microbial life in the sea is extremely diverse, including members of all three domains of life (archaea, bacteria and eukarya) as well as viruses (most authorities do not consider viruses as living organisms). The members of these groups or taxa are distinct in terms of their morphology, physiology and phylogeny and fall into both prokaryotic and eukaryotic domains. Considering the adaptability of microorganisms to grow and survive under varied physico-chemical conditions and their contribution in maintaining the balance in ecosystems, it is pertinent to catalogue their diversity as it exists. The inability to visualize them with the naked eye precludes effective classification.

The diversity of microbial communities varies within habitats as much as between habitats. Marine microbial habitats can be classifi based on i) presence of other organisms (Symbiotic, Free living and Biofi ii) proximity to ocean surface or sediments (Euphotic: 0-150 m; Mesopelagic: 150 – 1000 m; Bathopelagic : >1000m; Benthos : sediments) and also based on iii) concentration of nutrients and required growth substances (Oligotrophic, Mesotrophic, Eutrophic). However, interfaces tend to be hotspots of diversity and biological activity. Marine microbial habitats at interfaces include the air-water, water-sediment, water ice, and host macroorganism-water interfaces. The sub-millimeter scale of physical and chemical variability in these habitats poses a serious challenge to studying interface habitats in detail. The fact that variations can even occur within a few millimetres, suggests that microbial diversity encompasses more than the documented evidence available. Hence, biogeography is gaining importance as a fi of study from microbial diversity point of interest. Due to the innately small size of the microorganisms, environmental complexity plays a major role in determining diversity. Spatial heterogeneity is likely to lead to the formation of many niches within a habitat. Recent tools like metagenomics aid in biogeography studies by providing information on nucleic acid sequence data, thereby directly identifying microorganisms. Therefore the phylogenetic information can be used to compare microbial diversity profi across habitats. Generally, diversity within a particular location and in a community is called alpha diversity. Beta diversity measures the community composition between two or more locations while gamma diversity applies to a region, across continents and biomes and is larger in size than that used for measuring alpha diversity.

Although microbial diversity is one of the difficult areas of marine biodiversity research, estimation of microbial diversity is required for understanding the biogeography, community assembly and ecological processes. The number of species has been a traditional measure of biodiversity in ecology and conservation, but the biodiversity of an area is much more than the 'species richness'. Diversity prediction can be made using statistical approaches that estimate species number from relatively small sample sizes. Hughes et al. (2001) noted that both rarefaction and richness estimators, which have been applied to microbial datasets, highlighted the utility of nonparametric estimators in predicting and comparing bacterial species number. Rarefaction and richness estimators rely on a species or operational taxonomic unit (OTU) definition. The limitations of this method are that OTUs are counted equivalently despite the fact that some may be highly divergent and phylogenetically unique, whereas others may be closely related and phylogenetically redundant. Recently, statistical analyses borrowed from population genetics and systematics have been employed and reviewed for use with microbial datasets to estimate species richness and phylogenetic diversity which do not rely on estimation of the frequency of different sequences. Reciprocal of Simpson's index ($1/D$), F-statistics (FST) and phylogenetic grouping of taxa (P tests) may be used as a measure of diversity, which has been widely used for ecological studies. These combined uses of species richness and diversity estimates provide information that enables deeper understanding of marine microbial diversity.

During the past two decades, molecular studies using the gene sequences that encode the small subunit rRNA (SSU rDNA) have revealed a wealth of new marine microorganisms that belong to the three domains of life viz., bacteria, archaea and the eukarya. The feeling among marine microbiologists is that of living through an age of discovery with no end in sight. However, most of this new diversity has not yet been described because pure cultures of the organisms behind the sequences are necessary to define a species. Recent studies of microbial diversity have produced spectacular discoveries of previously unknown microorganisms, many of which have major impacts on oceanic processes. Very large populations of picoplankton including diatoms, dinoflagellates, picoflagellates and cyanobacteria are the primary catalysts in carbon fixation, which orchestrate the cycling of nitrogen and form the base of the traditional marine food web. Heterotrophic SAR11 represents the dominant clade in communities of oceansurface bacterioplankton while nonphotosynthetic protists of unknown diversity control the size of picoplankton populations and regulate the supply of nutrients into the ocean's food webs. Communities of Bacteria, Archaea, and Protists account for greater than 90% of oceanic biomass and 98% of primary production. Archaea includes unusual microorganisms which grow under extreme environments and differs from bacteria due to lack of peptidoglycan. Both these domains collectively play a significant role in the marine environment.

Modern technologies (molecular techniques and automated fluorescence cell sorting) have demonstrated the great abundance and diversity of microbial life forms in the oceans, and DNA sequencing of environmental genomes (metagenomics) provides evidence of hitherto unrecognized physiological categories among the planktonic microbes. Since 99% of the microbial population is considered to be uncultivable, metagenomics assumes importance. Molecular techniques have identified SAR11 as a dominant clade in communities of ocean-surface bacterioplankton. Bacteria in the SAR11 clade (Pelagibacteraceae) make up roughly one in three cells at the ocean's surface. Overall, SAR11 bacteria are estimated to make up between a quarter and a half of all prokaryotic cells in the ocean. SAR11 bacteria are classified as alphaproteobacteria, and include the highly abundant marine species Candidatus Pelagibacter ubique.

Sequence information (eg. 16S rRNA sequences, genome sequences, metagenomes) as well as rRNA targeted probes (eg. Fluorescent In Situ Hybridization – FISH, which allows a visual inspection of phylogenetic groups of cells in a natural sample) have helped to discover many new groups of bacteria. Marine microbial genomics ranging from the study of the genomes of model organisms to the wealth of meta-omics approaches (e.g. metagenomics, metatranscriptomics and metaproteomics) has proven to be very successful to target the second basic ecological question on the role of microbes in the marine environment. Advanced technologies, developed in the recent past, promise to revolutionize the way that we characterize, identify, and study microbial communities. The most advanced tools that microbial ecologists can use for the study of microbial communities include innovative microbial ecological DNA microarrays such as PhyloChip and GeoChip that have been developed for investigating the composition and function of microbial communities. Single Cell Genomics approach, which can be used for obtaining genomes from uncultured phyla, enables the amplification and sequencing of DNA from single cells obtained directly from environmental samples and is promising to revolutionize microbiology.

Microbial Taxonomy

Currently, a polyphasic approach is used to define a microbial species using phenotypic and genotypic properties. Whenever a new taxon is proposed, it is essential that the organism be isolated in pure culture and its characteristic features be tested under standard conditions. All the strains within a species must show similar phenotypes. A designated type strain of a species constitutes the reference specimen for that species. If the 16S rDNA sequences of organisms are ≤ 98.5% identical, they are members of different species. Uncultured microbes cannot be assigned to a definite species since their phenotype is not known; however, they can be assigned a 'Candidatus' designation provided their 16S rRNA sequence subscribes to the principles of identity with known species. A concept applying to a taxon lower than that of the strain is the ecotype – those microorganisms that occupy an ecological niche and are adapted to the conditions of that niche.

Numerical Taxonomy

In numerical taxonomy many (50 to 200) biochemical, morphological, and cultural characteristics, as well as susceptibilities to antibiotics and inorganic compounds, are used to determine the degree of similarity between organisms. The coefficient of similarity or percentage of similarity between strains (where strain indicates a single isolate from a specimen) is then calculated. A dendrogram or a similarity matrix is constructed that joins individual strains into groups and places one group with other groups on the basis of their percentage of similarity. While 16S rDNA sequences have attracted attention in recent times as sole means of bringing out the uniqueness of a species; numerical taxonomy (based on phenotypic traits of a large number of species) compares favourably with that of genotypic data and, indeed, is in alignment with the modern taxonomy.

Polyphasic Taxonomy

In polyphasic approach of microbial taxonomy, phenotypic, genotypic and phylogenetic information are described as accurately as possible. The phenotypic information comes from the colony characteristics, cell type, cell wall-type, pigmentation patterns, proteins and other chemotaxonomic markers while genotypic features are derived from the nucleic acids (DNA/RNA). Phylogenetic information is obtained from studying sequence similarities of the 16S rRNA or 23S rRNA genes in case of bacteria and 18S rRNA in case of fungi. Many types of molecules are used for delineating and describing a taxon; some are mandatory (16S rRNA genes, phenotypes, chemotaxonomy) while others are optional (amino-acid sequencing of certain protein products, DNA-DNA hybridization), unless required for appropriate description. DNA barcoding approach is gaining popularity for assessing microbial diversity. Though only limited datasets (especially for eukaryotic microbes) are currently available, the scenario is improving due to faster and cheaper sequencing methods.

Factors that Impact Marine Microbial Diversity

Nearly every measurable physical, chemical, and biotic variable in the marine environment has been found to alter microbial diversity. The major variables affecting the marine microbial diversity include, turbulance, light, temperature, nutreints, salinity, pH, UV and solar influx, surfaces and interfaces, redox potencial, metals as well as presence of macroorganisms such as invertebates and macroalgae. For the most part, the extent to which each of these influences actually operates in the environment and the contexts in which they are important remain to be determined. Climate change, which will be felt by marine microbial communities as changes in ocean temperatures, will undoubtedly alter the diversity of communities in unforeseen ways. Climate change should be considered a major top-down controller of microbial communities. Pollution, including nitrogen inputs due to anthropogenic activities, also impact marine microbial diversity. Anthropogenic nitrogen inputs to the oceans now comprise about half the total nitrogen inputs to the oceans, a circumstance that has resulted in vast dead zones in coastal areas and an increased incidence of harmful algal blooms.

Marine Microbes and Bioactive Compounds

The marine environment is emerging as a 'gold mine' for novel bioactive compounds. Marine de-rived natural products present an enormous range of novel chemical structures and provide an interesting and challenging blueprint for creating new entities via synthetic chemistry. Marine invertebrates and plants, in particular, represent an environment rich in microorganisms that produce compounds with bioactive properties including antibacterial, antifungal, antiviral, anti-cancer, antifouling and antibiofilm activities. However, only 1% of these microorganisms can be isolated using traditional culture techniques, which has been a major bottleneck when mining the marine environment for novel bioactive molecules. Marine microorganisms have immense poten-tial for providing services and products for human society which is not exploited to any significant extent. There is a world of microorganisms in the marine environment to discover, understand and put to good use.

Chapter 6

Marine Biodiversity Conservation

The conservation of biological and physical marine resources and ecosystem functions falls under marine biodiversity conservation. An understanding of marine biodiversity conservation follows from a study of marine biodiversity, the factors influencing biodiversity and the strategies that can be implemented to conserve it. This chapter covers all such vital aspects for a detailed understanding of this subject.

Marine Biodiversity

Marine biodiversity includes coastal and marine plant and animal species, their genetic variety, the habitats and ecosystems they form part of, and the ecological processes that support all of these.

The Earth is home to an estimated 10 million species. The largest of these are divided by biologists into 3 main kingdoms: fungi, plants and animals. The animal kingdom is divided into a further 33 distinct groups (or phyla).

Humans belong to the phylum called chordates, which includes all mammals, fishes, reptiles and birds. Other common phyla of the animal kingdom include the arthropods (insects and crustacea: crabs, shrimps and lobsters) and molluscs (snails, squid, octopus, cockles and mussels).

There are 11 phyla existing in terrestrial environments and 28 phyla living in the sea, of which 15 are exclusively marine. Examples of exclusively marine phyla include the echinoderms (starfish and their relatives), ctenophores (comb jellies), hemichordates (acorn worms) and the echiurans (trumpet worms).

The marine environment therefore includes a far greater diversity of animal groups than the terrestrial environment, which is not surprising since living organisms first appeared in the seas several hundred millions years before life on land evolved. Whether in the sea or on land, most plant and animal species are grouped into assemblages or communities characteristic of recognizable habitats.

The eastern African coast, for example, includes mangrove, seagrass and coral reef habitats. Each of these requires specific environmental conditions for its development. In the case of the mangrove habitat, shelter from wave action and soft mud or sand are the basic conditions that allow the community to flourish.

The combination of habitats forms the marine ecosystem. This ecosystem, the various habitats, communities and species they comprise, constitute the marine biodiversity of the eastern African region.

Factors Affecting Marine Biodiversity

Biodiversity loss has become one of the greatest environmental concerns of the last century, owing to increasing pressure on the environment by humans combined with the realisation that our activities can seriously threaten the future sustainability of marine species and ecosystems. Marine biodiversity in Europe is threatened by the fact that many of the goods and services provided by marine ecosystems are exploited in a non-sustainable way. In some cases, marine ecosystems are threatened to the extent that their structure and function is being jeopardised.

Threats to marine biodiversity have widespread social, economic, and biological consequences, the combination of which could threaten our own existence, including:

- Economic losses through unemployment and reduced productivity
- Dramatic reductions in the numbers of many popular edible fish and shellfish
- Extinction of species that might be useful in developing new medicines
- Reduced ability of ecosystems to respond to disaster, both natural (floods) and man-made (pollution)
- Accelerated global climate change
- Social and political instability

The most serious threats to marine biodiversity are:

Overexploitation

Overexploitation or overfishing is the removal of marine living resources to levels that can not sustain viable populations. Ultimately, overexploitation can lead to resource depletion and put a number of threatened and endangered species at risk for extinction.

Overexploitation of fish stocks is a major
threat to marine biodiversity in Europe

A greater variety of species at a higher trophic level is exploited in the sea than on land: humans exploit over 400 species as food resources from the marine environment; whereas on land only tens of species are harvested for commercial use. Exploitation of marine biodiversity is also far less managed than on land and amounts to the hunter-gatherers stage that humans abandoned on land over 10,000 years ago, yet exploitation technology is becoming so advanced that many marine species are threatened to extinction. Insufficient consideration has been given to the unexpected and unpredictable long-term effects that such primitive food-gathering practices engender.

The Problem

The exponential growth in human population experienced in last decades has lead to an overexploitation of marine living resources to meet growing demand for food. Worldwide, fishing fleets are two to three times as large as needed to take present day catches of fish and other marine species and as what our oceans can sustainably support. The use of modern techniques to facilitate harvesting, transport and storage has accelerated this trend. According to the United Nations Food and Agriculture Organization (FAO) over 25% of all the world's fish stocks are either overexploited or depleted and 52% are fully exploited. Thus a total of almost 80% of the world's fisheries are fully to overexploited, depleted, or in a state of collapse. Although, these estimates are considered rather conservative. Recently, a study showed that 29% of fish and seafood species have collapsed (i.e. their catch has declined by 90%) and are projected to collapse within by 2048, unless inmediate action is taken. Worldwide about 90% of the stocks of large predatory fish stocks are already collapsed.

Overexploitations do not only affect open ocean or pelagic ecosystems, but also coastal and intertidal areas. For example, intertidal limpets in Hawaii (Cellana spp.), the Azores, Madeira and Canaries (Patella spp.) have all shown declines, and in the case of the Azores, dramatic population crashes owing to food gathering.

Overexploitation Effects

All fishing activities, if not conducted in a sustainable non-destructive manner, can lead to overexploitation of marine living resources. Overexploitation of marine resources has major impacts on marine systems as a whole, but target species are generally the most impacted.

Fishing effect can be divided into: direct effect and indirect effects. Direct effects are related to target species and by-catch species.

Direct Effects

When recruitment of target species is relatively high, the average size of individuals is affected because larger individuals tend to be harvested and populations display signs of growth overfishing. When adult populations are heavily exploited the number and size of the adult population (spawning biomass) is reduced to a point that it has not the reproductive capacity to replenish itself, leading to recruitment overfishing. Direct effects of fishing also include physical disturbance by fishing gear than can cause scraping, scouring and resuspension of the substratum. The effects vary according to the gears used and the habitats fished.

Trawling for demersal species is having a major effect on the habitat for species other than target species. It has been estimated that all of the sea bed of the North Sea is trawled over at least twice per year and the gear is getting heavier over time. Trawls have destroyed long-lived species of molluscs and echinoderms in the North Sea. Since these species play important functional roles in biogeochemical cycling the consequences may be far-reaching.

Tuna fisheries represent a typical case were the growth overfishing of stocks can be generated by technological progress

Indirect Effect

Fishing not only has direct effects on target populations but also results in indirect effects such as effect of "goast fishing", trophic cascading effects and food web-competion.

Trophic cascading effects has been observed when top-level predators are removed resulting in indirect effects throughout the ecosystem. On many temperate reefs shifts from macroalgae-dominated habitats to habitats grazed by sea urchins, termed 'urchin barrens', have been linked to the over-harvesting of top predators. Perhaps the best-known example of this is the interaction between sea otters, sea urchins and kelp. The importance of the sea otter-urchin-kelp trophic cascade was demonstrated after sea otters were wiped out by harvesting for their fur, allowing their prey, sea urchins, to overgraze kelps and dominate many benthic ecosystems. After the repopulation of areas by otters, kelp and its associated communities became much more abundant.

Indirect effects of fishing include "Ghost fishing" from fishing nets left or lost in the ocean by fishermen

Ghost fishing result from fishing nets that have been left or lost in the ocean by fishermen. These nets, often nearly invisible in the dim light, can be left tangled on a rocky reef or drifting in the open sea. They can entangle fish, dolphins, sea turtles, sharks, dugongs, crocodiles, seabirds, crabs, and other creatures, including the occasional human diver. Acting as designed, the nets restrict movement, causing starvation, laceration and infection, and — in those that need to return to the surface to breathe — suffocation.

Conservation Tools

Conservation tools and techniques for preserving marine biodiversity conservation tend to combine theoretical disciplines, such as population and community marine biology, with practical conservation strategies, such as setting up marine reserves or marine protected areas (MPAs). Other techniques include developing sustainable fisheries, which involves establishing fishing quotas and restoring the populations of endangered species through artificial means.

Another focus of conservation effort is to halt human activities that are detrimental to the marine environment through policies or legislations, at an international, European and/or regional level.

Education of the general public about conservation issues is key in the process of conservation of the marine environment. There are many marine conservation organizations throughout the world that focus on funding conservation efforts, educating the public and stakeholders, and lobbying for conservation law and policy.

Marine Reserves and Marine Protected Areas

Very little (less than 1%) of the earth's oceans are protected, compared to 12% of the land surface. Highly-protected Marine Reserves are areas of the sea where human disturbances are minimized, allowing to maintain the natural biodiversity or, more often, to recover it to a more natural state. In Europe there are very few Marine Reserves, they are small and almost all are in the Mediterranean. While, there are many Marine Protected Areas (MPAs) in Europe, these areas only have some extra regulations or planning procedures, but they do not fully protect marine biodiversity. Inside Marine Reserves all extractive and potentially-disturbing human activities are prohibited. Marine Reserves are of crucial importance to science and education, essential for conservation, useful in resource management and necessary for reversing the effects of overexploitation. The benefits of marine reserves has been shown in New Zealand where twenty marine reserves have been created over the last 25 years. For example, inside Marine Reserves the abundance of species of ecological and commercial importance, such as the snapper Pagurus auratus and the spiny lobster Jasus edwardsii are almost 9 and 4 times, respectively, more abundant inside reserves than in adjacent unprotected areas. Furthermore, the mean body size of these animals is significantly large inside reserves than in outside. There is an urgent need for a representative, replicated, networked and sustainable system of highly protected Marine Reserves.

Habitat destruction and fragmentation is a process that describes the emergences of discontinuities (fragmentation) or the loss (destruction) of the environment inhabited by an organism. Marine ecosystems are experiencing high rates of habitat loss and degradation, and these processes are considered as the most critical threat to marine biodiversity. It is estimated that every day between 1960 and 1995, a kilometer of coastline was developed, causing permanent losses of valuable habitats, such as coastal wetlands, seagrass meadows and rocky shores. Approximately 20% of the world's coral reefs were lost and an additional 20% degraded in the last several decades of the twentieth century, and approximately 35% of mangrove area was lost during this time.

Habitat can be defined as predominant features that create structural complexity in the environment, such as plants (e.g., seagrass meadows, kelp forests), or animals (e.g. Sabellaria reefs, burrowing fauna in sandflats).

Marine habitats in Europe, such as kelp forests, support
a high diversity of marine life and are threatened by human activity

Habitat Loss and Biodiversity

Researchers have identified three major components of habitat loss that relates to species diversity has been identified:

1) The loss of resident species.

There is a unique plant and animal diversity living in close association with specifics habitats or habitat forming species. For example, invertebrate assemblages associated with mussel beds, epiphytic and epibenthic assemblages communities living in kelp forests or fish communities associated with coral reefs. It has been predicted that the consequences of habitat loss would probably be much more profund than than the loss of individual species, because of the ecological interactions between species leading to a chain of impacts.

The loss of habitats can lead to the formation of
barren areas where only few species are able to survive

2) The loss of food resources.

Most biogenic habitats are highly productive compared to simpler habitats. They produce large amounts of nutrients and organic matter that can be directly used by other organisms as food resources. The habitat loss also implies the loss of these food resources having a negative effect in the survival of other species and the productivity of individual species or communities, with more profound effects that are likely to propagate along food chains.

3) The loss of ecosystem functions provided by the habitat.

Structurally complex habitats provide a wide range of ecosystem functions to the environment including food and refuge provision for other species, trapping sediment, modifying light and hydrodynamic conditions, providing resilience to the system. When the habitats are lost this functions are lost with them. For example, the replacement of macroalgal canopies by turfs affects sediment dynamics on rocky coasts, where fronds prevents accumulation of sediments while turfs tend to trap sediments even on exposed coasts

Effects of Habitat Loss on Biodiversity

Coral reefs are an important structurally complex habitat
that provide food resources and refuge to many fish species

Habitat loss has been generally associated to drastic declines in overall abundance and diversity of marine organisms. For example, in the Wadden Sea, the destruction of biogenic habitats has caused the regional extinction of at least 26 species during the past 2000 years. Similarly, the loss of seagrass meadows results in a reduction in the number of species and abundance of fishes. Generally, the environmental changes associated with the destruction of natural habitats promote the arrival and colonization of opportunistic species that can benefit from conditions in disturbed condition. One example is the expansion of opportunistic green ephemeral algae or turf forming algae following the destruction of removal of canopy forming species, such as kelp. Other species that can favor from disturbed habitats condition are alien species. Once alien species are established they can contribute further to reduce local diversity by interacting with native species.

Human Activities as Driver of Habitat Loss

Some anthropogenic activities responsible for habitat destruction include the construction of coastal protection, land reclamation, aggregate (sand and gravel) extraction, recreation and developments including ports, harbours and industries. Additionally, the growing number of tourists presents a significant threat to many coastal habitats in Europe, which can disturb by trampling or direct harvesting. In offshore waters, exploration and development of oil and gas activities threaten marine habitats, mainly with discharges of oil and other pollutants. Physical damage to marine habitats can result from fishing activities such as bottom trawling. Deep-water trawlers use heavy rock-hopping equipment, which has been reported to cause long-term to seabed habitats such as cold-water coral reefs in Norwegian, Scottish and Irish waters.

Implications for Conservation and Management

Habitat loss often leads species to get endangered or threatened, and eventually extincted leading to significant loss of overall diversity and changes in ecosystem functioning. Conservation efforts of coastal and marine habitats have been driven in part by the effects of habitat loss on declines in species richness. However, looking at the effect of habitats loss on species richness is not sufficient.

Conservation efforts of coastal and marine habitats have been driven in part by the effects of habitat loss on declines in species richness. However, looking at the effect of habitats loss on species richness is not sufficient. Conservation efforts must consider the effects habitat loss and fragmentation on all components of species diversity and the ecosystem functioning and services provided by habitat:

No take marine reserves are one of the most effective management tools to protect endangered habitats

- regulating services such as shoreline buffering from storms

- provisioning services such as fish production

- cultural services such as tourism

- supporting services such as primary production.

There is an urgent need for the implementation of long-term and large scale monitoring programs of changes to marine habitats and species distributions. This monitoring should be coupled with studies designed to treat management actions that result in habitat loss as large-scale experiments. For example, the use of marine protected areas (MPAs) and marine reserves as tools to underpinned the relationships between the functionality of habitats and the distribution and abundance of target taxa. According to the Global Biodiversity Assessment the most effective way to conserve biodiversity, by almost any reckoning is to prevent the conversion or degradation of habitat.

There is an ongoing debate among conservationist biologist about whether is preferable to protect several already fragmented patches of habitats or a single large area, often referred as the SLOSS (Single Large or Several Small) debate.

One important aspect that also needs to be considered is habitat restoration. On land there is a long tradition of restoring habitats, such as mining waste tips. There are some examples of habitat restoration in the marine environment, such as the well-publicized clean-up of the River Thames in the UK where salmon can now be found in London. The developing science of restoration ecology should be a part of a strategy for conservation of coastal biodiversity.

An alien or non-native species is one that has been intentionally or accidentally transported and released into an environment outside of its historic or resident geographical range or habitat. Such species are described as 'invasive' if they are ecologically and/or economically harmful. Invasive species can dramatically change the structure and function of marine ecosystems by changing biodiversity and eliminating vital components of the food chain.

Functional Integrity of Species Commuities

Even though all seas are interconnected, there are certain dispersion barriers such as salinity, temperature gradient or ocean currents that keep local communities distinct. Two identical habitats can have different species communities with different types of interactions contributing to the global richness of biodiversity. Indigenous, or native species are those living within their natural range (past or present) including the area that it can reach and occupy using its natural dispersal system. By contrast, introduced species are transported either intentionally or accidentally by human-mediated vectors into habitats outside their native range. These species are also termed as alien, exotic, invasive, foreign, non-native, immigrant, neobiota, naturalized, or non-indigenous. Some biota cannot be sufficiently proved to be neither native nor exotic and these are termed cryptogenic species.

Through the evolutionary fitness processes, species adjust to each other and adapt to available resources by occupying different ecological niches within communities. Virtually all resources are utilized optimally and all available niches are filled, maximizing the biodiversity value. Various factors influence the functional integrity of a community. If changes are occurring gradually over a long timescale species have enough time to adapt and fill available niches. In turn more rapid shifts create niche openings and this has been identified as the main prerequisite of species invasion. Non-native species are found primarily in disturbed areas, such as harbours, bays, estuaries and semi-enclosed seas where the communities are weekend by various types of pollution.

Vectors of Introduction

Whether deliberately or accidentally, people have been transporting whole range of organisms, breaking natural distribution boundaries and interfering with community structures. The unwanted hitchhikers are usually either well hidden or too small to be noticed – for the entire lifespan or just for its part.

For most coastal species the open ocean environment is inhospitable, preventing them from spreading into habitats similar to their own but located elsewhere. Distances separating such habitats might be too long to overcome either through their active swimming abilities or passive floating in water currents. Mechanisms by which humanity aids introduction of exotic species are called vectors of introduction and these are chiefly associated with shipment activities, marine aquaculture or ornamental species trade. Other vectors include the international transport and sale of live marine bait, live seafood, and live organisms for research and education.

- Shipment

  Organisms can travel either attached to the submerged part of the vessels hull or contained within their ballast waters. Ballast water has been extensively used since 1870s and certain species are able to complete their life cycle and breed in ballast waters, meaning that they can be translocated far away from their native range. Studies on ballast tanks found more than 1,000 taxa ranging from phytoplankton to small fish up to 15 cm in length.

- Mariculture

  There has been a growing interest in the search for fish, shellfish, and plant species whose biology was well known and which either already had achieved or could achieve success in cultivation. Depending on the type of mariculture, the organisms can be either allowed to establish in the wilderness (intentional species introduction) or kept in enclosures from which they occasionally escape (accidental release) either directly or through their dispersal system. Very often those new introduced species carried along parasitic organisms that could later establish themselves in indigenous species. Since the latter have not developed any defence strategy against non-indigenous parasites, they could be greatly impacted.

- Ornamental species trade

  Saltwater species are a rapidly growing sector of the aquarium industry. Species can be released by their owners either accidentally or deliberately when no longer wanted. This vector is responsible for one of the major invasions launched up to date, namely the release of Caulerpa taxifolia from Oceanographic Museum of Monaco.

- Indirect introduction

  Occasionally, species introduction can result not from their physical re-location but by offering a way for its dispersal to areas that it wouldn't be able to reach if the conditions were not changed. Such opportunity is offered by alteration of hydrological regimes, like canal and reservoir construction.

- Secondary introduction

  If species are introduced and become established in a new geographic setting, they might continue to spread by both natural and anthropogenic mechanisms, colonizing habitats that they wouldn't be able to reach without the initial, human-mediated translocation. Such introduction is termed a secondary introduction.

Very rarely we can connect an invasion event with one particular vector. Instead it is more common to assess the probability of a given vector being responsible. In British waters accidental release associated with mariculture has been identified as the main vector with other important vectors being fouling, ballast water and deliberate commercial introduction.

From Establishment to Invasion

Not all translocated species become established in new environments. If the population is relatively small it will be vulnerable to stochastic threats such as demographic or genetic drift. It is hard to predict what is the minimum viable population size. A generally accepted rule is that 50 individuals are needed to prevent excessive inbreeding and a minimum population of 500 hundred is required to keep a sufficiently high level of heterozigosity. Yet, those numbers may very greatly between different taxa and even if the population is large enough to prevent loss of alleles, it still might be prone to environmental threats such as poor food or oxygen availability, impacting recruitment success or juveniles development, or catastrophic events.

If the founder population is large enough to overcome stochastic threats and manages to establish itself in a new environment, it will join the network of interactions within the receiving environment. Alien species have evolved in different environments and it is hard to predict what their interactions with indigenous biota will be. Not all introduced species can be termed as invasive. Some of them may well coexist with native species and share the resources. But every now and then an introduced species becomes an invader and impacts the host community, with the ultimate result being destabilization of the system and possibly extinction of native species.

Organisms have been carried around hidden in dry ballast, attached to hulls or buried inside them for millennia and it is very likely that what we now consider cosmopolitan species are simply early introductions and that species such as ship-boring isopod (Sphaeroma terebrans), Asian seasquirt (Stylea plicata), giant kelp (Macrocystis pyrifera), mussels (Mytilus galloprovincialis and Mytilus edulis) and European periwinkle (Littorina litorea) can be possible early introductions. This in turn raises the issue of naturalisation: How much time is needed to fully incorporate a population in a community and bring the latter to equilibrium? How many interactions have to be established to call an introduced species integrated? Carlton also questioned the quality of our understanding of present marine communities. If we assumed a rather timid scenario that between 1,500 and 1,800 only three species were introduced elsewhere each year, we would have ended up with a number of 1,000 species that might have spread before humanity gained a general knowledge of biogeography, taxonomy and ecology. As a result, these species can be today considered as cosmopolitan. It might have a great impact on our understanding of marine ecology and community equilibrium in particular.

Impacts

Introduced colonial sea squirt (Didemnum sp.) fouling other organisms, including the commercially important blue mussel Mytilus edulis in Irish Sea.

- Competition

  Introduced species might have evolved in the presence of much more aggressive competitors than the ones present in the receiving environment and they might be extremely successful in colonizing new areas. The most common competition is for resources, whether it is food, solar energy or space. In California bay the native shore crab (Hemigrapsus oregonensis) declined in mean abundance by 10 times as a consequence of the European green crab (Carcinus maenas) introduction and competition for food (Nutricola sp.).

- Diffuse competition

  The impact on native organisms can be greater through diffuse competition. A population might resist competition along one axis if the resource is plentiful, but its realized niche can virtually disappear in the presence of several species competing along many axes.

- Positive indirect interactions

  Even with a stable or decreasing rate of new introductions, it can be assumed that systems will become increasingly destabilized through positive indirect interactions and diffuse competition, as more and more interactions within the community become altered. Even early, non-invasive introductions may become a nuisance due to a change induced by another introduced species. It is possible for a new alien to transform an older exotic species into an invader.

- Predation and herbivory

  Some organisms are more effective competitors for resources than others and have the potential to dominate the area, excluding other species from the access to viable resources. In such cases organisms feeding on it (predators or herbivores) act as a biocontrol agent keeping their abundance low and maintain a high level of stability and productivity. If those predators or herbivores are introduced to a new environment, local community members might be lacking the evolutionary-driven ability to resist them. On the other hand if a species limited in number in its native range by biocontrol agents is introduced to a new environment, it might thrive and outcompete indigenous species.

- Parasitism

  Very often introduced organisms are accompanied by parasitic species. Mass transfer of large numbers of animals and plants without inspection, quarantine, or other management procedures has led to the simultaneous introduction of pathogenic or parasitic agents.

- Genetic impact

  If transferred species are closely related to native counterparts, they may hybridize affecting the original gene pool. Introduced or hybridized individuals may prove to be more successful in mating than the indigenous organisms and the original genetic diversity might be lost.

- Multiple negative interactions

  Two species may interact with one another in a several ways. A species confronted by one-way interactions can compromise and trade-off one component of fitness for the sake of the survival. If it is forced to compromise too many components it may be pushed into an extinction vortex.

The Problem

Consequences of human-induced altering of species composition are sometimes detrimental to local communities and their magnitude may range from limitation or exclusion of single species or to destabilization of the whole system and species introduction has been identified as one of the main causes of species extinction. Even with a stable or decreasing rate of new introductions, it can be assumed that systems will become increasingly destabilized through direct and indirect interactions and diffuse competition as more and more interactions within the community become altered. In areas that have already been heavily invaded, reducing the numbers of new introductions may not be a sufficient strategy. In addition to preventing new introductions, it may be necessary to mitigate the impacts of exotic species that have already become established. Given the large number of alien species already present, there is a high potential for positive interactions to produce many future management problems.

At least 50 non-native species have entered the Black Sea in the last century and some have been invasive. For instance, the comb jelly-fish Mnemiopsis leidyi was the primary cause of collapse of the fisheries in the area in the early 1990s. Over 100 non-native species have been recorded in the NE Atlantic, mainly in the North Sea, the Celtic Sea, and the Bay of Biscay and along the Iberian coast. Impacts of invasive species vary in different regions and sometimes are rather localized. Over the past twenty years, the number of alien species transported into the Baltic has increased and poses a significant threat to the region given its naturally low species diversity. In North America, some 300 non-indigenous species of invertebrates and algae have been established in marine and estuarine waters. Rates of invasion among most microscopic organisms such as bacteria are still unreported and it appears that their potential of dispersal by human-mediated vector is very significant.

The comb jelly Mnemiopsis leidyi (native to western Atlantic) feeds on eggs and larvae of pelagic fish and caused a dramatic drop in fish populations in the Black Sea by competing for the same food sources and eating the young and eggs.

Even without human-induced climate change, the biodiversity and biogeography of species is continuously changing (seasonal and yearly changes). Consequently, long term monitoring is necessary in order to evaluate these processes. The marine systems however may become more dynamic and variable due to climate change.

Europe may be less threatened by sea-level rise than many developing country regions. However, coastal ecosystems do appear to be threatened, especially enclosed seas such as the Baltic, the Mediterranean and the Black Sea. These seas have only small and primarily east-west orientated movement corridors, which may restrict northward displacement of organisms in these areas.

Effects on Primary Production

Higher temperatures and enhanced stratification could affect the productivity of phytoplankton. A number of models predict an increase in global primary production of between 1% and 8% by 2050, when compared to pre-industrial times.

Because phytoplankton is an important basis of the marine food web, any change in the timing, abundance or species composition of the phytoplankton will have an effect on the whole food web.

Effects on the Recruitment Process

The population dynamics of a lot of marine vertebrates and fish are driven by recruitment processes. The recruitment of cold temperate species is often synchronized with seasonal production cycles of phytoplankton. Increasing sea water temperatures may advance the timing of reproduction of these fish species; this may result in a mismatch with their food source (phytoplankton) (match/mismatch hypothesis). A change in recruitment success will lead to shifts in species composition.

Atlantic cod Gadus morhua

The Atlantic cod (Gadus morhua) recruitment in the North Sea, in the past 40 years, was influenced by changes at the base of the food web (bottom-up-control), induced by the rise of temperature. Cod recruitment decreased from the mid-1980s, coincident with unfavorable changes in the plankton ecosystem.

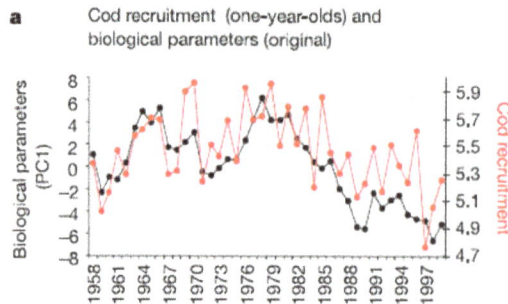

Long-term changes (1958-1999) in the plankton index (in black), and long-term changes (1959-2000) in cod recruitment (in red, decimal logarithm).

Effects on the Biogeography

The species movement in a warming area is towards the poles in general. Since global warming accelerated in the late 1980s, pole ward advances of southern species and retreats of northern species have been recorded in zooplankton, fish and benthic species.

The species distribution is not always northwards. For example when the stock of the main prey of the harp seals in the Barents Sea collapsed, these seals migrated southwards along the coast of Norway and into the North Sea in search of food.

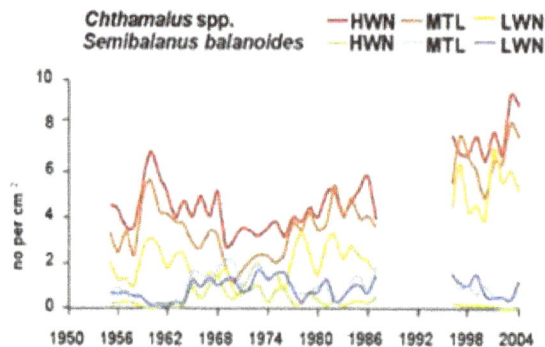

Long-term changes in Semibalanus balanoides and Chthamalus spp. for several shores at different submersion levels on the south coast of Devon and Cornwall.

Example: Study of barnacles in the Celtic-Biscay shelf

The Celtic-Biscay shelf was liable to warming in the 1930; in the 1960 there was a switch back to colder. Changes in species assemblages were described on rocky shores, using barnacles as a sensitive indicator of wider changes in marine life. These showed switches between warm water barnacles in the 1950s (Chthamalus spp.) to greater dominance by the cold-water barnacle Semibalanus balanoides in the 1960s and 1970s. On rocky shores warm water barnacles now exceed the levels found in the 1950s.

There are also examples from the North Sea and the Baltic Sea.

Effects on the Phenological Relationships and Community Structure

The response to climate changes differs between the species, inducing a decoupling of phenological relationships (relative timing of life cycle events). The decoupling may affect the community structure and food webs by altering the interactions between a species and its competitors, mutualists, predators, prey or pathogens. For example, in the case of seabirds, chick diet composition during development is likely to be an important mechanistic link between climate variability and the observed decline in seabird populations.

A study in Kongsfjorden (79 °N) concluded that changing temperatures have a direct and very immediate influence on the species composition of plankton as well as on the biodiversity. The response of benthic organisms to rising temperature is slower and less drastic compared with planktonic organisms. Planktonic organisms may be a better indicator of global warming-driven changes than the benthic organisms. However benthic fauna is a good indicator of slow changes, especially when observed over a long period of time.

Effects on the Establishment of Invasive Species

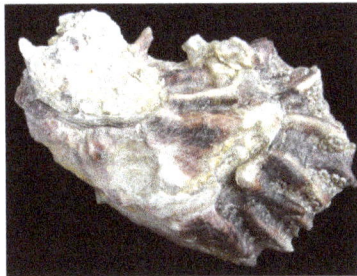

Pacific oyster Crassostrea gigas

The establishment of non-indigenous species can be accelerated by rapid warming. For example, the recent warming has accelerated the adaptation of the Pacific oyster (Crassostrea gigas) on the local circumstances in the Netherlands and the UK.

Effects on Biogeochemical Cycles

In the past 200 years the oceans have absorbed approximately half of the CO_2 produced by fossil fuel burning and cement production. Calculations indicate that this uptake of CO_2 has led to a reduction of the pH of surface seawater (acidification) of 0.1 units; this is an equivalent to a 30% increase in the concentration of hydrogen ions. If the CO_2 emissions from human activities rise on current trends then the average pH of the oceans could fall by 0.5 units by the year 2100.

This fall in pH may have a huge impact on marine organisms, in particular to calcifying organisms such as most mollusks, corals, echinoderms, foraminifera and calcareous algae. Seawater has to be supersaturated with calcium and carbonate ions to ensure that once the biogenic calcareous structures are formed, it does not dissolve. Lower pH reduces the carbonate saturation of the seawater, making calcification harder.

There is also a difference in vulnerability between the groups of organisms: corals and a group of mollusks (pteropods) precipitate aragonite; coccolithophores and foraminifers produce the less soluble calcite. Furthermore differs the function (e.g. metabolic or structural function) from the carbonate between the different groups and this is coupled to the sensitivity for acidification. It is likely as CO_2 levels increases, changes of species composition will occur because of the different responses of the species. An altered species composition may have a huge effect on the global carbon cycle.

Effects at the Level of Physiological Responses to Temperature Rise

Temperature rise for the oceans as a whole is likely to be about 1 to 2 °C, in the next decades. Exceptions are semi enclosed marine lagoons and shallow bays that may mirror the atmospheric temperature rise as well. Additional, most aquatic sandy-shore animals are adapted to rapid changes in temperature and they seldom experience temperatures close to their upper tolerance limits. Further, sandy-beach animals are capable of burrowing and of escaping below the sand if conditions at the surface become hostile. This is in contrast to sessile organisms such as corals and mangroves that are unable to keep up with the higher temperature level.

Ocean Pollution

There are various ways by which pollution enters the ocean. Some of them are:

1. Sewage

Pollution can enter the ocean directly. Sewage or polluting substances flow through sewage, rivers, or drainages directly into the ocean. This is often how minerals and substances from mining camps find their way into the ocean.

The release of other chemical nutrients into the ocean's ecosystem leads to reductions in oxygen levels, the decay of plant life, a severe decline in the quality of the sea water itself. As a result, all levels of oceanic life, plants and animals, are highly affected.

2. Toxic Chemicals From Industries

Industrial and agricultural wastes are another most common form of wastes that are directly discharged into the oceans, resulting in ocean pollution. The dumping of toxic liquids in the ocean directly affects the marine life as they are considered hazardous and secondly, they raise the temperature of the ocean, known as thermal pollution, as the temperature of these liquids is quite high. Animals and plants that cannot survive at higher temperatures eventually perish.

3. Land Runoff

Land runoff is another source of pollution in the ocean. This occurs when water infiltrates the soil to its maximum extent and the excess water from rain, flooding or melting flows over the land and into the ocean. Often times, this water picks up man-made, harmful contaminants that pollute the ocean, including fertilizers, petroleum, pesticides and other forms of soil contaminants. Fertilizers and waste from land animals and humans can be a huge detriment to the ocean by creating dead zones.

4. Large Scale Oil Spills

Ship pollution is a huge source of ocean pollution, the most devastating effect of which is oil spills. Crude oil lasts for years in the sea and is extremely toxic to marine life, often suffocating marine animals to death once it entraps them. Crude oil is also extremely difficult to clean up, unfortunately meaning that when it is split; it is usually there to stay.

In addition, many ships lose thousands of crates each year due to storms, emergencies, and accidents. This causes noise pollution (excessive, unexpected noise that interrupts the balance of life, most often caused by modes of transportation), excessive algae, and ballast water. Often times, other species can also invade an ecosystem and do harm to it by interrupting the life cycles of other organisms, causing a clash of nature that has already been damaged by the overflow of pollution.

5. Ocean Mining

Ocean mining in the deep sea is yet another source of ocean pollution. Ocean mining sites drilling for silver, gold, copper, cobalt and zinc create sulfide deposits up to three and a half thousand meters down in to the ocean. While we have yet the gathering of scientific evidence to fully explain the harsh environmental impacts of deep sea mining, we do have a general idea that deep sea mining causes damage to the lowest levels of the ocean and increase the toxicity of the region. This permanent damage dealt also causes leaking, corrosion and oil spills that only drastically further hinder the ecosystem of the region.

6. Littering

Pollution from the atmosphere is, believe it or not, a huge source of ocean pollution. This occurs when objects that are far inland are blown by the wind over long distances and end up in the ocean. These objects can be anything from natural things like dust and sand, to man-made objects such as debris and trash. Most debris, especially plastic debris, cannot decompose and remains suspended in the oceans current for years.

Animals can become snagged on the plastic or mistake it for food, slowly killing them over a long period of time. Animals who are most often the victims of plastic debris include turtles, dolphins, fish, sharks, crabs, sea birds, and crocodiles.

In addition, the temperature of the ocean is highly affected by carbon dioxide and climate changes, which impacts primarily the ecosystems and fish communities that live in the ocean. In particular, the rising levels of CO_2 acidify the ocean in the form of acid rain. Even though the ocean can absorb carbon dioxide that originates from the atmosphere, the carbon dioxide levels are steadily increasing and the ocean's absorbing mechanisms, due to the rising of the ocean's temperatures, are unable to keep up with the pace.

Effects of Ocean Pollution

1. Effect of Toxic Wastes on Marine Animals

Oil spill is dangerous to marine life in several ways. The oil spilled in the ocean could get on to the gills and feathers of marine animals, which makes it difficult for them to move or fly properly or feed their children. The long term effect on marine life can include cancer, failure in the reproductive system, behavioral changes, and even death.

2. Disruption to the Cycle of Coral Reefs

Oil spill floats on the surface of water and prevents sunlight from reaching to marine plants and affects in the process of photosynthesis. Skin irritation, eye irritation, lung and liver problems can impact marine life over long period of time.

3. Depletes Oxygen Content in Water

Most of the debris in the ocean does not decompose and remain in the ocean for years. It uses oxygen as it degrades. As a result of this, oxygen levels go down. When oxygen levels go down, the chances of survival of marine animals like whales, turtles, sharks, dolphins, penguins for long time also goes down.

4. Failure in the Reproductive System of Sea Animals

Industrial and agricultural wastes include various poisonous chemicals that are considered hazardous for marine life. Chemicals from pesticides can accumulate in the fatty tissue of animals, leading to failure in their reproductive system.

5. Effect on Food Chain

Chemicals used in industries and agriculture get washed into the rivers and from there are carried into the oceans. These chemicals do not get dissolved and sink at the bottom of the ocean. Small animals ingest these chemicals and are later eaten by large animals, which then affects the whole food chain.

6. Affects Human Health

Animals from impacted food chain are then eaten by humans which affects their health as toxins from these contaminated animals gets deposited in the tissues of people and can lead to cancer, birth defects or long term health problems.

Strategies to Conserve Marine Biodiversity

The ocean is downstream of everything, so all of our actions, no matter where we live, effect the ocean and the marine life it holds. Those who live right on the coastline will have the most direct impact on the ocean, but even if you live far inland, there are many things you can do that will help marine life.

Eat Eco-friendly Fish

Our food choices have a huge impact on the environment — from the actual items we eat to the way they are harvested, processed, and shipped. Going vegan is better for the environment, but you can take small steps in the right direction by eating eco-friendly fish and eating local as much as possible. If you eat seafood, eat fish that is harvested in a sustainable way, which means eating species that have a healthy population, and whose harvest minimizes bycatch and impacts on the environment.

Limit your use of Plastics, Disposables and Single-use Projects

Plastic bag floating twenty miles offshore. Blue Ocean Society

Have you heard of the Great Pacific Garbage Patch? That is a name coined to describe the huge amounts of plastic bits and other marine debris floating in the North Pacific Subtropical Gyre, one of five major ocean gyres in the world. Sadly, all the gyres seem to have their own garbage patch.

What is the problem? Plastic stays around for hundreds of years can be a hazard to wildlife and leaches toxins into the environment. The solution? Stop using so much plastic. Buy things with less packaging, don't use disposable items and use reusable bags instead of plastic ones wherever possible.

Stop the Problem of Ocean Acidification

Mussels (Mytilus edulis) feeding, Ireland

Global warming has been a hot topic in the ocean world, and it is because of ocean acidification, known as 'the other global warming problem.' As the acidity of the oceans increases, it will have devastating impacts on marine life, including plankton, corals and shellfish, and the animals that eat them.

But you can do something about this problem right now — reduce global warming by taking simple steps that will likely save money in the long run — drive less, walk more, use less electricity and water — you know the drill. Lessening your "carbon footprint" will help marine life miles from your home. The idea of an acidic ocean is scary, but we can bring the oceans to a more healthy state with some easy changes in our behavior.

Be Energy-efficient

Polar Bears Sleeping

Along with the tip above, reduce your energy consumption and carbon output wherever possible. This includes simple things like turning off the lights or TV when you're not in a room and driving in a way that increases your fuel efficiency.

Participate in a Cleanup

Trash in the environment can be hazardous to marine life, and people too. Help clean up a local beach, park or roadway and pick up that litter before it gets into the marine environment. Even trash hundreds of miles from the ocean can eventually float or blow into the ocean. The International Coastal Cleanup is one way to get involved — that is a cleanup that occurs each September. You can also contact your local coastal zone management office or department of environmental protection to see if they organize any cleanups.

Volunteers at a beach cleanup

Never Release Balloons

Balloons may look pretty when you release them, but they are a danger to wildlife such as sea turtles, who can swallow them accidentally, mistake them for food, or get tangled up in their strings. After your party, pop the balloons and throw them in the trash instead of releasing them.

Dispose of Fishing Line Responsibly

California sea lion

Monofilament fishing line takes about 600 years to degrade. If left in the ocean, it can provide an entangling web that threatens whales, pinnipeds and fish (including the fish people like to catch and eat). Never discard your fishing line into the water — dispose of it responsibly by recycling it if you can, or into the garbage.

View Marine Life Responsibly

Two humpback whales lunge-feeding near a whale
watch boat as passengers look on in awe.

If you're going to be viewing marine life, take steps to do so responsibly. Watch marine life from
the shore by going tide pooling. Take steps to plan a whale watch, diving trip or other excursions
with a responsible operator. Think twice about "swim with dolphins" programs, which may not be
good for dolphins and could even be harmful to people.

Volunteer or Work with Marine Life

Scuba diver and a whale shark (Rhincodon typus)
in the Indian Ocean, Ningaloo Reef, Australia.

Maybe you work with marine life already or are studying to become a marine biologist. Even if
working with marine life isn't your career path, you can volunteer. If you live near the coast, volunteer opportunities may be easy to find.

Buy Ocean-friendly Gifts

Give a gift that will help marine life. Memberships and honorary donations to non-profit organizations that protect marine life can be a great gift. How about a basket of environmentally-friendly
bath or cleaning products, or a gift certificate for a whale watch or snorkeling trip? And when you
wrap your gift - be creative and use something that can be re-used, like a beach towel, dish towel,
basket or gift bag.

Marine Protected Areas

A marine protected area (MPA) is essentially a space in the ocean where human activities are more strictly regulated than the surrounding waters - similar to parks we have on land. These places are given special protections for natural or historic marine resources by local, state, territorial, native, regional, or national authorities. Authorities differ substantially from nation to nation.

A clearly defined geographical space, recognised, dedicated and managed, through legal or other effective means, to achieve the long-term conservation of nature with associated ecosystem services and cultural values.

Types of MPA

There are many kinds of marine protected areas which can have a wide range of conservation objectives. Such objectives can include:

Ecological objectives:

- To ensure the long-term viability and maintaining the genetic diversity of marine species and systems;

- To protect depleted, threatened, rare or endangered species and populations;

- To preserve habitats considered critical for the survival and/or lifecycles of species, including economically important species;

- To prevent outside activities from detrimentally affecting the marine protected areas;

Human objectives:

- To provide for the continued welfare of people affected by the creation of marine protected areas;

- To preserve, protect, and manage historical and cultural sites and natural aesthetic values of marine and estuarine areas, for present and future generations;

- To facilitate the interpretation of marine and estuarine systems for the purposes of conservation, education and tourism;

- To accommodate with appropriate management systems a broad spectrum of human activities compatible with the primary goal in marine and estuarine settings; and

- To provide for research and training, and for monitoring the environmental effect of human activities, including the direct and indirect effects of development and adjacent land-use practices.

Some people confuse marine reserves, where extraction of any resources is prohibited (no-take), as the only type of MPA. MPAs may include marine reserves, as well as other zones in which partial protection is afforded (seasonal closures, catch limits, etc.). Many MPAs are multiple-use areas, where a variety of uses are allowed. For example, there are many different kinds of MPAs in U.S. waters including national parks, wildlife refuges, monuments and marine sanctuaries, fisheries closures, critical habitat, habitat areas of particular concern, state parks, conservation areas, estuarine reserves and preserves, and numerous others. While a few sites exist as no-take marine reserves, the vast majority of MPAs, both in terms of numbers and area, are open for fishing, diving, boating, and other recreational and commercial uses.

The types of human activities that are regulated, and the strictness of the regulations, is therefore largely dependent upon the objectives of the MPA.

Types of Regulation and MPA Management Techniques

MPA Community Surveillance, Bamboung

There are a range of management techniques that MPA managers can use. These techniques can be broadly categorized into ways of prohibiting and limiting activity:

Prohibition: Absolute prohibition of access to a prescribed area is the simplest form of regulation. It is a form of control that establishes a clear yes/no basis – if a person is found in the area, he has violated the regulation. Prohibition of certain activities within a prescribed area is another prohibitive technique. For example, if fishing is prohibited in a specific area and a person is caught fishing there, he is in violation.

Limitations: Both terrestrial and marine protected areas around the world often allow some level of human activity, especially if it involves recreation, nature appreciation, education, or research. The management challenge is to design and enforce measures that limit allowed human activities to levels that do not cause harmful or unacceptable impacts. Limitations are also more challenging than prohibitions – they are more complex for area users to understand and may be more difficult for managers to enforce. However, limiting rather than prohibiting activities in an area is usually more acceptable to area users and may be more easily implemented. Limitation by spatial control involves regulating activities specifically to a part or parts of the MPA.

- Zonal management: Spatial control of activities.

- Temporal control: Management changes over time, such as a closed fishing season. For example, this may be used to protect spawning areas for fish or breeding habitats for seabirds.

- Equipment restriction: Regulation of the use of equipment or technology that is efficient for its purpose in the short term but damaging to resources in the long term (e.g., trawl restrictions).

- Quotas: Setting limits on the allowable harvest with the goal of leaving enough of the resource to replenish itself. Quotas are most commonly applied towards fishing.

- Licenses or permits: Issuing permission, through official documentation, for a person or people to engage in specific activities in the MPA. Licenses and/or permits can be issued based on skill, resource allocation, or other characteristics.

MPAs are just one of many marine resource management tools. MPAs primarily regulate human activities by segregating them spatially. MPAs alone cannot address problems such as pollution, climate change, or overfishing. Other management strategies are needed to complement marine reserves. They are most effective when used in conjunction with other management measures.

Constructive public engagement in MPA planning is vital to achieving conservation goals; both in establishing sites and in ensuring their effective long-term stewardship. In many cases, different authorities and agencies seek public input on the design, location, and management plan for new MPAs or no-take areas within existing MPAs. Approaches used to acquire stakeholder input varies widely depending on agency-specific requirements, policies, timelines, and other constraints. Public engagement in these very different planning processes ranges from sustained substantive involvement over several years, to more limited participation focused mainly on commenting on internally generated preliminary plans.

The optimum size, number, and location of MPAs are determined by the management goals of a particular area. There are three basic designs that are most commonly used and discussed: a small single area, a large single area, or a network of areas. A small area may be used to protect a unique habitat, a site-specific life cycle event (such as spawning aggregation that occurs in a single area), or a unique shipwreck. A large single area may be used to protect species nursery grounds, representative habitat from either fishing pressure or destruction of habitat, or a large collection of historic vessels. A network of MPAs may be used to protect habitats needed for the diversity of life stages common among marine species to ensure that larval transport occurs throughout an entire region.

Most management systems for MPAs will use a variety of management approaches to achieve the MPA objectives.

References

- Habitat-destruction-and-fragmentation: marbef.org, Retrieved 16 May 2018
- Non-native-species-invasions: marbef.org, Retrieved 31 March 2018
- Effects-of-global-climate-change-on-European-marine-biodiversity: marbef.org, Retrieved 12 May 2018
- Causes-and-effects-of-ocean-pollution: conserve-energy-future.com, Retrieved 12 July 2018
- Easy-ways-to-help-marine-life-2291549: thoughtco.com, Retrieved 22 June 2018

Permissions

Index

www.ingramcontent.com/pod-product-compliance
Lightning Source LLC
Chambersburg PA
CBHW082021190326
41458CB00010B/3234